D1232174

# Sophie's Diary

By

**Dora Musielak**

authorHOUSE®

*AuthorHouse™*
*1663 Liberty Drive, Suite 200*
*Bloomington, IN 47403*
*www.authorhouse.com*
*Phone: 1-800-839-8640*

*First published by Authorhouse  02/12/08*

*ISBN: 1-4184-0811-5 (e)*
*ISBN: 1-4184-0812-3 (sc)*

*Library of Congress Control Number:  2003099685*

*Printed in the United States of America*
*Bloomington, Indiana*

*This book is printed on acid-free paper..*

*Paris, France*
*Year 1789*

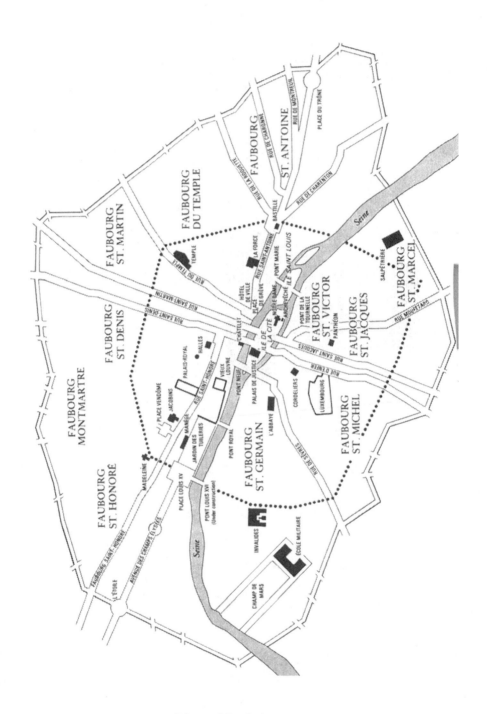

Map of Paris in 1789.

*A long time ago, in a far away land, a sage woman raised three boys to be honest and smart. When her husband died, he bequeathed her one milking cow, and thirty-five cows more to be divided among their three sons. According to the wishes of the father, the oldest was to receive half the number of cows, the second would get a third of the total, and the youngest would receive a ninth of all.*

*The brothers began to quarrel amongst themselves, as they did not know how to divide the herd. They attempted different ways to divide it, and the more they tried, the angrier they became. Every time one of the brothers proposed a way to divide the animals, the other two would shout that it was unfair.*

*None of the suggested divisions was acceptable to all. When the mother saw them squabbling, she asked why. The oldest son replied: "Half of thirty-five is seventeen and one-half, yet if the third and the ninth of thirty-five are not exact quantities either, how could we divide the thirty-five cows amongst the three?" The mother smiled, saying the division was simple. She promised to give each their share of the herd and assured her sons it would be fair.*

*Do you know how the wise woman divided the herd?*

CHARGE

3

# 1

## Awakening

My NAME IS SOPHIE. Today is my thirteenth birthday and my parents surprised me with some wonderful gifts. My father gave me a lap desk of glossy mahogany wood. It has a compartment to keep paper and pens, and a secret drawer that can only be opened by pressing a hidden tab underneath. When I unlocked it, I found a mathematical riddle and a note from Papa, challenging me to solve it. He gives me a year, but I hope to find the answer sooner than that. Oh, I am so excited!

My mother gave me a porcelain inkwell, a stack of linen paper, sealing wax, a miniature burner, six bottles of ink in different colors, and a set of quills. These are the best birthday presents ever! I promised my parents to write often to improve my penmanship. Maman says writing neatly and correctly is very important for a refined young lady. This gave me the idea of writing a *journal intime*, a diary where I can record my innermost thoughts and feelings.

I love my father. He calls me *ma petite élève*, my little pupil. He taught me arithmetic when I was five. I still remember the excitement I felt when Papa taught me the numbers. I used to think that the arithmetic operations he taught me were magic tricks to produce new numbers. Just for fun, I spent hours combining numbers to create others. Now I help Papa with the more complicated calculations he needs in his business.

I have two sisters, Marie-Angelique and Marie-Madeleine. Angelique is ten years old, a pretty and vivacious girl who loves to chitchat; exactly the opposite of me. Madeleine is almost nineteen years old. She resembles our mother and, like her, she is romantic and sweet. I am not like either of them. My sisters like to dress in beautiful gowns, style their hair, and sing and dance. I don't like any of that.

Angelique calls me "peculiar." She teases me all the time because I like to be alone and would rather read books than socialize. I do not care whether my hair is curled, or if my dresses are fashionable. I do not think about these things like she does. But I am fond of her, just as I am of my mother and Madeleine. I especially love my father with all my heart; *mon cher père*, he understands me better than anyone.

We live in a townhouse on rue St Denis not far from the Seine River. It is one of the most important avenues in Paris. Kings and princesses enter the city through St Denis. This neighborhood is lively. From my bedroom window I see the imposing buildings of the Conciergerie and the Palais de Justice. I can hear the bustle of the people on the Pont au Change, the bridge over the river that we cross going to Mass at Notre Dame.

The church bells are tolling the eleventh hour. I must stop writing now and go to bed. But I still feel giddy, thrilled with happiness and tenderness towards my parents for giving me these wonderful gifts for writing. From now on, I shall use ink and paper to express my own ideas and thoughts, just as my parents suggest.


Sunday—April 5, 1789


What a lovely day. A bit cold but the sun broke through the clouds and the last traces of ice lingering on the rooftops began to melt. Paris awakes from the long slumber of winter, so harsh this year. In January, it was so cold that the water in the house froze. We were sick with sore throats and coughing. My mother was worse, as she succumbed to pneumonia. Thankfully, she recovered and now she looks radiant.

My mother is a very traditional, old-fashioned lady who insists that my sisters and I learn the graces of society. She teaches us good manners, just as her mother taught her. She believes that young women must learn domestic arts to become good wives. I won't, for I will never get married! Madeleine, however, hopes to wed Monsieur Lherbette very soon. I like him. Sometimes he converses with me in the foyer while waiting for my sister. He is always polite and asks me questions about the books I am reading.

Our family life is simple. We begin every day with prayer, and then my father retires to his library to review his business accounts. We dress and get ready for the day. After breakfast my father goes to work, and my sisters and I join Mother in her sitting room. We spend the morning sewing, knitting, and learning to needlepoint. When the weather is nice, we go for a stroll in the gardens. Angelique plays with other children while I read next to Maman, sitting on a bench by my favorite tree. Once a week, Monsieur Chevalier comes to give me piano lessons. I find these lessons tiresome. Not that I do not like music; on the contrary, I love it, but I would rather listen to others play. I have no talent, especially compared with Madeleine. Even Angelique plays the piano much better than me.

Papa comes home to eat lunch at one o'clock. In the afternoon my mother instructs me in penmanship. This is a lesson I enjoy because Maman chooses passages from important books that I copy to help me learn the correct manner of writing. Every Friday after dinner, my mother reads us classical plays, sometimes novels and poetry. Angelique and Madeleine recite whatever love poems they find. When Papa is home, he talks about philosophy and history, his favorite subjects. After supper, my mother and Madeleine play cards with Papa. Every Tuesday night Papa and his colleagues meet in his study to discuss politics. I read or play Chinese checkers by myself, but if Papa is not busy he plays chess with me. He is very good, but I usually win the match.

I dislike embroidering, but Maman insists I learn to do it. I'm so clumsy with handicrafts and can never finish even the simplest design. At night, Millie tiptoes into my bedroom and finishes the embroidery for me. I hope Maman does not find out or she'll be upset with me and reprimand Millie. Millie's job is to help Madame Morrell with the household chores, run errands, and look after my little sister. Monsieur and Madame Morrell are Millie's parents. The three have worked for my family for many years. Monsieur Morrell is Papa's valet and coachman, and Madame Morrell cooks, runs the household, and assists Maman with her coiffure.

My favorite place in the house is my father's library, where I read every day. I also like to be there because I can listen to the conversations Papa has with his friends. They talk about politics and that usually leads to discussions of philosophy, literature, and other interesting topics. The best part is that the gentlemen don't seem to mind that I am there, sitting quietly in a corner. Or perhaps they don't notice me.

My father is a very intelligent man and I admire him dearly. Lately, Papa and his colleagues have discussed the deterioration of the economy, which affects the poorest citizens of French. Papa talks often about the shortage of goods, the rising unemployment and the increase in taxes; he says the workers are planning protests, thus aggravating the situation. Some people blame King Louis XVI for the dwindling economy. Papa believes that the heavy taxation on the peasants and poor workers is unfair. He support social equality but admits is difficult to change a society like ours, with aristocrats and top church leaders having all the wealth and social privileges. Maman defends the king, but she can't excuse the social injustices.

Oh, I hear the clock chime the midnight hour. I must go to sleep now or I won't be up on time for morning prayers.

Friday—April 10, 1789

Last night I roamed my father's library looking for something to read. I found an interesting book titled *History of Mathematics,* authored by Jean-Étienne Montucla. The history is fascinating and illuminating, as it describes how mathematics evolved. The first chapters deal with the development of mathematics in antiquity.

Among other things, I read about Archimedes, a Greek scholar who lived in a Greek colony called Syracuse hundreds of years ago. Archimedes was an exceptional man. The author describes Archimedes as introverted and absentminded, just like me. Archimedes would become so involved in his study that he paid no attention to unimportant things around him. He'd draw geometric figures on any available surface, including the dust on the ground or in the ashes from extinguished fire. The author describes how Archimedes solved a very interesting problem that goes like this. One day the king of Syracuse gave an artisan a certain amount of gold to make him a crown. The finished crown was beautiful, but for some reason the monarch suspected that the craftsman had stolen some of the gold and replaced it with silver to make up the required amount needed to craft the crown. The king could not prove the artisan's deception, so he asked Archimedes for advice.

Archimedes had to find a rational method to confirm or refute the king's suspicion. It happened by serendipity. One day, when Archimedes entered his bathtub, he observed that the amount of water that overflowed the tub was proportional to the weight of his body submerged in the water. That gave him the idea to solve the problem of the king's crown. Archimedes was so excited that he ran naked through the streets of Syracuse shouting like a lunatic *"Eureka! Eureka!"* which means, "I have found it! I have found it!" Archimedes found the Law of Buoyancy as he described it in his book *On Floating Bodies*. This law states that a floating object always pushes aside an amount of water equal to the weight of the object. Archimedes also discovered that a nonfloating object—one that sank—pushed aside an amount of water equal to the object's volume. The great scholar went on to prove that the crown was indeed made with two different metals.

Archimedes died tragically while doing mathematics. He was so absorbed in his studies, that Archimedes was unaware of the Romans invading Syracuse. As it was his custom, that day Archimedes was drawing figures in the dust when a Roman soldier stepped on them. Archimedes was very upset and told him, "Do not disturb my circles!" The soldier was so enraged by the old man's outburst that he pulled out his sword and killed him.

After reading this part of the book, I could not sleep. I kept imagining Archimedes, a person so consumed by a mathematical problem that he did not notice what was going on around him. A strange feeling overcame me, a need to experience the same kind of passion that Archimedes must have felt.

Then I understood my desire to know more of this science. Mathematics is not just about the simple calculations that ordinary merchants perform every day. Oh, I am happy to know basic arithmetic, to add and multiply numbers, but it is insufficient. No, it is not enough that I know the simple operations that Papa taught me. I want to learn much more to become a great mathematician, just like Archimedes.

There are many concepts that I wish to learn and understand, like the number $\pi$, which is prominent in the story of Archimedes. I need to know what $\pi$ means and why it is so important. Reading a book of mathematics, full of symbols and equations, is like reading a book written in another language. I want to speak the language of mathematics!

How I wish I had someone to teach me. Antoine-August, the son of one of Papa's friend, is studying mathematics with a tutor. This morning I asked Maman if I could have a teacher too, but she stopped me, looking at me as if I was insane. Frowning, Maman stated that girls do not study mathematics and asked me to expunge those thoughts from my mind. Why? I am a girl and I like mathematics very much. Why should I stop thinking about it?

Wednesday—April 15, 1789

Now I know what $\pi$ means. First, $\pi$ is a letter in the Greek alphabet used by mathematicians to represent an important number. As a number, $\pi$ represents the ratio of the circumference to the diameter of the circle. It is such a unique number that there is no fraction equal to it, and its value is the same for all circles, regardless of their size.

To prove it, I could draw a circle and measure its circumference with a piece of string, which I would place on the line that represents the circle, exactly once around. The length of the string should correspond to the circumference, C, of the circle. I would measure the diameter, D, of the circle, which is the length from any point on the circle straight to its center to another point on the opposite side of the circle. Then I would divide the circumference of the circle by its diameter. This ratio C/D is equal to $\pi$. Interestingly, if I draw smaller or bigger circles, the result is the same: C/D is *always approximately* equal to 3.1415. In other words, π is a constant with the same approximate value for all circles.

But if I could measure and divide perfectly, could I get an exact value of π? Based on what I know, the answer is no! The number π is *not* an exact ratio. The approximated value of π has been known for thousands of years, but its value was obtained by measurement, therefore it was not very accurate. One day, Archimedes proposed a mathematical procedure that used geometry instead of direct measurement. He wrote *The Measurement of a Circle*, a book describing how he took a polygon with 96 sides and inscribed a circle inside the polygon, and with his geometric analysis estimated the value of π. Archimedes stated that "π is a number between $3\frac{10}{71}$ and $3\frac{1}{7}$."

Now mathematicians write π = 3.14159.... The dots after the nine mean that the exact value of π is unknown. One can keep on refining the calculation, thus getting more digits on the right side of the decimal point, but so far nobody knows how many more.

I've read two full chapters of the *History of Mathematics* and learned about another ancient mathematician named Pythagoras. He was a Greek philosopher who also studied astronomy and developed music theory. Pythagoras moved to Croton, a city in southern Italy (a Greek colony) and started a religious and philosophical school. This great man had many followers called the Pythagoreans, scholars who also studied mathematics; after the master died, they continued the schools for many more years. Pythagoras believed that most things could be understood through numbers. Not only did this lead to developments of new mathematical concepts, but also to those of science and philosophy as well. Much of the mathematics of today is due to Pythagoras and his followers. That's why I intend to investigate more about it.

Sunday—April 26, 1789

This afternoon Angelique taunted me and called me *gauche*. I did not say anything, but her incessant teasing hurts me. We came

11

from church and were having Sunday lunch; everybody was in a fine mood. Then Mother mentioned the invitation from Madame LeBlanc to attend a dinner party next week. I do not like these social functions where everybody talks and laughs, telling each other silly stories. I never know what to say, and I feel awkward and out of place. It doesn't help that Angelique makes fun of my shyness. The worst part is that, like it or not, I have to go to Mme LeBlanc's soirée. Mother thinks it is important for my social development, that it will teach me to behave like a lady.

I would rather stay home and read books, to learn about extraordinary people like Pythagoras. Numbers fascinated him, thinking that they possessed mystical powers. Amazingly, he found relationships between numbers and things that at first appear unrelated to mathematics, like music! Pythagoras discovered the relationship between the length of a taut string of a harp and the sound it produces when the string is plucked. He experimented with different lengths and found that if the string is shortened to ½ its original measurement, then the tone produced is one octave higher than the original tone. That's how Pythagoras discovered musical intervals. Pythagoras influenced the development of mathematics to this day. He founded a school for men and women to study mathematics and science. The biography of Pythagoras suggests that his wife was also a mathematician, but her name is not given. Why is that? Does it mean her accomplishments were not as important as her husband's? Or that she was shy like me and didn't like to show off her smarts? Who knows. After all, these events happened centuries ago!

Tuesday—April 28, 1789

I always thought Paris was the richest city in Europe, but now I am not so sure. Many people are poor and wander the streets begging. And food is scarce. Millie told Maman this morning that people started a fight when the baker ran out of baguettes. To ensure we get enough bread, Millie goes to the bakery very early and waits in a long line for hours. The people that come late don't get any and the poorest cannot afford it. According to Papa, laborers earn twenty-five sous a day, and now the four-pound loaf costs fourteen or fifteen sous! Millie added that some workers were exchanging their shirts for a loaf of bread and, in one case, a woman even tried to exchange her corset.

A demonstration erupted at the market yesterday. Millie said it all began with a speech against social injustice. Then a man began to shout, inciting the shopkeepers to revolt, and soon several others followed. The mob chanted, "Death to the rich! Death to the aristocrats!" and were growing more rowdy by the minute. The violence of the protests is escalating. It's scary.

Later in the afternoon, another riot broke out in the faubourg Saint-Antoine. The angry workers, acting on rumors of wage cuts, destroyed the wallpaper factory of Monsieur Réveillon, a business friend of Papa. Many people were killed when the government troops opened fire at the mutineers. The workers, afraid of losing their jobs or not earning enough to afford bread, started with a protest. One thing led to another, and the protest boiled over in violence. The fire also consumed the family's house. Their great library, with hundreds of valuable books, was burned to ashes and now there is nothing left. The only good thing is that M. Réveillon and his family escaped unharmed.

I continue reading. The *History of Mathematics* is one of those precious gems in Papa's library that are too hard to put down. I finished another fascinating chapter, full of anecdotes about ancient scholars. I read about Hypatia of Alexandria, the first woman scholar of considerable merit. Hypatia was the daughter of the mathematician Theon, who lived around 400 A.D. This remarkable woman was the head of a school in Alexandria where she taught the ideas of the Greek philosopher Plato. Scholars of that era esteemed Hypatia for her intelligence and vast knowledge. Hypatia helped her father write commentaries on important mathematical books, such as Euclid's *Elements*. Hypatia's greatest achievement was her commentaries on the *Arithmetica* of Diophantus, one of the most important books for mathematicians.

Hypatia died most tragically, massacred by people who felt threatened by her intelligence and knowledge. A rumor started that Hypatia was responsible for a conflict that arose between two important spiritual leaders. It is not clear why she was blamed and why some men disliked Hypatia so much. One day the men seized Hypatia on the street, beat her to death, dragged her body, mutilated her flesh with sharp tiles, and burned her remains.

Poor Hypatia; what a horrible way to die! How inhuman, how cruel! I tremble just imagining her nightmare.

Saturday—May 2, 1789

My father was elected deputy of the Constituent Assembly. In this capacity, he will represent the citizens of the Third Estate before the court. Next week he will go to Versailles to attend a meeting with the king and discuss the economic problems. Father knows of the country's huge debt and is aware of its effects on society. In fact, he suggests several social reforms to restore balance. Many people blame King Louis XVI and Queen Marie Antoinette for their economic problems. But the debt was already immense when they ascended the throne. Regardless of who is at fault, the tax increase is creating serious social and economical conflicts.

My father acknowledges that taxes are not distributed equally among the citizens of France. That is the reason why the workers and peasants are so unhappy. It's an unfair situation for them since the clergy and aristocrats, who represent the First and Second Estates, are exempt from taxation. That leaves only the Third Estate—the rest of the people—to pay the debt. Any reasonable person would see an obvious solution: force the nobles and the clergy to pay their share of taxes. A solution like this one would strengthen the ideals of equality in our society. Father agrees, but he says the privileged aristocrats are fiercely opposed, and they will do anything to avoid losing their wealth and property.

I never thought about it before, but now it all becomes clear. I begin to understand what the French philosophers of the Enlightenment meant about social inequality. Why are there social classes? When did all this begin? Long ago, when the primitive societies formed, someone assumed power; but how? Who dictated that one person would control the lives of many others? Right now, Louis XVI rules France because he inherited the throne from his grandfather, Louis XV. I can name the kings of France in the recent past, but I cannot tell who the first king was. How did the monarchies of the world begin? It is not fair that, under this social structure, only a few individuals have power and wealth, and the remaining citizens have none. My father has suggested many times that a monarchy is unnecessary. He thinks it is possible for a country to exist without a king. Can France become a republic ruled by the people?

Monday—May 4, 1789

Today was a historical day for France. It began with a great parade through the streets of Paris to inaugurate the dialogue between the Estates General and Louis XVI to solve the economic problems. This exchange of ideas is the first of its kind in the history of France. The king recognizes the needs of his subjects and is willing to listen to their demands. We, like all Parisians, went to watch the parade. But it was more meaningful for us, as Father was in the procession with the other deputies of the Third Estate.

Louis XVI and Marie Antoinette led the solemn march. Princes and princesses, wearing the most beautiful attire, followed the royal family. The ladies wore stunning brocade and taffeta gowns, and their hair was piled high, adorned with colorful feathers. They were resplendent, with their jeweled tiaras shinning against the sun. The queen seemed a bit forlorn. Maybe she still mourns the baby she lost two years ago, or maybe she is worried for the Dauphin of France, her sickly seven-year-old son. But she walked with grace and dignity and smiled to her subjects. The bishops and other leaders of the Church of France marched behind the aristocrats, wearing their richest robes of silk and lace. The royal falconers strode with hooded birds attached to their wrists.

As the parade made its way, the people lining the streets shouted, "*Vive le roi, vive la reine!*" "Long live the king, long live the queen!" A few men bellowed, "Long live Duc d'Orleans!" to acknowledge the king's cousin who ardently supports the ideals for social equality.

And last, the deputies of the Third Estate walked somberly behind, all dressed in simple black outfits that were in stark contrast to the rich costumes of the clergy and the nobility. Even without such regalia Father looked so handsome! With his virile walk and stature, Papa was a commanding presence among the six hundred deputies. Angelique waved at him excitedly and yelled, "*Vive le tiers état!*" I doubt he saw us among the immense crowd.

The parade culminated with a high Mass at the Church of Saint Louis but we didn't attend. We returned home, tired and hungry, but excited and so proud of Papa. As a deputy of the Third Estate he will get an opportunity to express his ideas for economic reform and equality.

Sunday—May 10, 1789

Some people believe that Babylonian mathematicians knew the Pythagorean Theorem hundreds of years before Pythagoras established it in the form we know it. The Pythagorean Theorem is a mathematical statement of great simplicity and so useful. It says that *in a right triangle, the square of the hypotenuse is equal to the sum of the squares of the legs.* I can write the theorem using an algebraic equation:

$$x^2 + y^2 = z^2$$

where $x$ and $y$ are the sides of the right triangle, and $z$ is the hypotenuse.

It is a simple, easy-to-solve equation, yet it is very powerful. The Pythagorean Theorem can be used to calculate areas, lengths, and heights. For example, suppose I have two points on a plane, and I want to know how far apart they are. I can either measure it directly, or I can measure the distance along the $x$-axis, measure the distance along the $y$-axis, and then use the Pythagorean Theorem to find the distance. In many situations it is easier to measure the $x$ and $y$ distances than the total distance between two points.

16

For example, suppose that I have a right triangle for which I know the length of the two short sides, say $x = 7$ and $y = 3$. Then I can calculate the value of the longest side of the triangle, or hypotenuse, by applying the Pythagorean Theorem. The square of the hypotenuse $z$ is equal to the sum of the squares of the two sides:

$$z^2 = 7^2 + 3^2$$
$$z^2 = 49 + 9$$
$$z^2 = 58$$
$$z = \text{square root of } 58$$

The square root of 58 is 7.61..., a number that has a decimal part with many digits.

Now, if $x = 3$, and $y = 4$, I get:

$$z^2 = 3^2 + 4^2$$
$$z^2 = 9 + 16$$
$$z^2 = 25$$
$$z = 5.$$

In this case, $5^2 = 3^2 + 4^2$ represents perfect squares.

The Pythagorean equation leads me to two types of numbers: perfect whole numbers, and fractional numbers. Now, what about negative numbers? Is it possible to get negative numbers? No, since a negative number multiplied by itself will yield a positive number.

I am sure there is much more to learn about this theorem. But it has to wait until tomorrow. The church bells tolled the eleventh hour a while ago. It must be near midnight!

Saturday—May 16, 1789

I am tired. We spent almost all day at the Palais Royal. This is the most popular place in Paris, especially during the weekend. It was crowded, with people strolling in the congested archways.

Women wearing the most extravagant hats strutted along, proudly displaying their fashionable dresses. We went to a café and found a table under the trees in the central courtyard. Maman ordered Austrian pastries and, while we enjoyed our delicious dessert, we watched a troupe of actors made up like marionettes. Gypsy women wearing colorful skirts also danced, and some were going around reading people's fortunes in their hands.

A young man suddenly jumped on a table nearby and began a speech, inciting citizens to fight for equality and justice. Some people stopped to listen, but many more ignored the eloquent orator. A few men distributed pamphlets with invitations to join groups to fight injustice. The message was clear: France needs to start a revolution.

At noon we watched the solar cannon. This is another popular diversion that attracts many onlookers. My sister Angelique was the first in line to see the cannon firing at midday, as she said, "to salute the sun." But she was disappointed that the cannon did not detonate again. I had to explain to her that it can only shoot once a day, when the rays of the sun strike the lens at the noon hour. I told her that the lens acts as a burning glass, which precisely focuses on the touchhold of the small cannon, and so it fires automatically, just once a day. I'm not sure Angelique understands that.

I have my own amusement with mathematics. I've been thinking about the number $\pi$. I know that $22/7$ is a good approximation to its value, and $355/113$ is a much better one. But there is no exact number, all I can write is $\pi = 3.14159....$ Most likely the number expansion on the right side of the decimal point is infinite. Maybe it is infinite because the circumference of the circle is an infinite number itself. This is what I think: since $\pi$ is equal to the ratio of the circumference to the diameter of a circle, $C/D$, and the circumference $C$ is the length of the line that represents the circle, and this length does not have a beginning or an end, I conclude that dividing something infinite must give an infinite value.

In addition to its relation to the circle, why else is this number so important? Perhaps $\pi$ is the solution to an equation that describes the universe. Especially, if the universe were round! Oh, I am sure that learning more mathematics can help me answer these and many other questions that burst in my mind.

A violent thunderstorm woke me up a few minutes ago. As the lightning parted the darkness of the night, I jumped out of bed. The room became fully illuminated as if a thousand candles were lit at once; but just as fast, it became pitch dark. It must be about three o'clock in the morning!

Few things scare me, but thunderbolts terrify me. I've never told this to anyone; who would understand? During a lightning storm I tremble like a scared child. I wish I could find refuge in my mother's arms, as I imagine the warmth of her tender embrace would soothe my fear; the sound of her beating heart would muffle the frightful sounds. But I am not a child anymore. I must confront my fears and be strong. I will just sit here and wait until the storm ends.

I know my fear of lightning is irrational. Irrational as in "not rational" or illogical. This reminds me of something. In my study of numbers, I found rational and irrational numbers, but in mathematics, "irrational" does not mean illogical. In fact, irrationals are some of the most amazingly beautiful numbers, although they are very rare. Irrational numbers are real numbers that cannot be represented as fractions. Irrational numbers result from dividing two real numbers, but they are not perfect ratios like 1/2, 12/3, 24/4, which are called rational numbers.

So, if the ratio of two real numbers results in a number with a decimal part that goes on and on without end, and the digits in the decimal part is randomly distributed, then the number is irrational. For example, $\pi$, which is approximately equal to 3.14159..., is an irrational number because its value results from dividing two real numbers (the circumference and the diameter) and is not exact; the digits on the right side of the decimal point go on; and their sequence does not repeat.

Hundreds of years ago, Pythagoras discovered that harmony in musical instruments is determined by the ratio of the length of the strings. He also thought that the planets orbit in harmony to ratios. Pythagoras considered ratios as the mystical concept that the Divine Mathematician used to create the universe and everything in it. Pythagoras rejected any number that did not conform to this ideal.

Thus, the Pythagorean mathematicians that followed his teachings also overlooked irrational numbers.

But irrational numbers do exist; an example is the square root of 2. This number is a fraction with a non-repeating decimal that goes on without end. This non-ratio later became known as an irrational number. The Pythagorean philosopher Hippasus used geometric methods to demonstrate the irrationality of $\sqrt{2}$. There is a legend that tells how Hippasus was thrown overboard a ship for studying irrational numbers and proving that the square root of two is irrational. When Hippasus announced his discovery, his outraged colleagues threw him overboard for going against the teachings of Pythagoras.

I don't think it really happened. All I know is that Pythagorean mathematicians did not work with irrational numbers. This is the only disappointing fact about the great mathematician of antiquity.

Thursday—May 28, 1789

I think about numbers all the time. I don't see numbers just as tools for adding or subtracting, as that would make mathematics too simple. I think about all types of numbers such as integers, rational and irrational numbers, primes, and perfect numbers. I want to learn what makes these numbers different and how to use them.

Integers are numbers that do not have a decimal part like 1, 13, 27, and so on. When I write 1, 2, 3, I say that these are consecutive integers. In general, consecutive integer numbers are integers $n_1$ and $n_2$, such that $n_2 - n_1 = 1$; i.e., $n_2$ follows immediately after $n_1$. So, given two consecutive numbers, one must be even and the other must be odd. Since the product of an even number and an odd number is always even, the product of two consecutive numbers, and, in fact, of any number of consecutive numbers, is always even.

An integer of the form $n = 2k$, where $k$ is an integer, results in even numbers, such as, –4, –2, 0, 2, 4, 6, 8, 10, ... The product of an even number and an odd number is always even, as can be seen by writing $(2k)(2m + 1) = 2[k(2m + 1)]$, which is divisible by 2, and hence is even.

A number that can be expressed as a fraction $p/q$, where $p$ and $q$ are integers and $q \neq 0$, is called a rational number with numerator $p$ and denominator $q$. When the fraction is divided out it becomes a number with a terminating or repeating decimal, such as $\frac{1}{2} = 0.5$, and $5/3 = 1.666666\ldots$.

Fractional numbers that are not rational are called irrational numbers. An irrational number cannot be expressed as a fraction $p/q$. In decimal form, irrational numbers do not repeat in a pattern or terminate; they go on and on, such as $\pi = 3.141592\ldots$ and $\sqrt{2} = 1.414213\ldots$.

And then there are primes, the most intriguing and fascinating among all numbers. A prime number is a positive integer $p > 1$ that has no positive integer divisors other than 1 and $p$ itself. More concisely, a prime number $p$ is a positive integer having exactly one positive divisor other than 1. For example, the only divisors of 13 are 1 and 13, making 13 a prime number, while the number 24 is not because 24 has divisors 1, 2, 3, 4, 6, 8, 12, and 24, corresponding to the factorization $(24 = 2^3 \cdot 3)$. Although the number 1 was considered a prime number, it is not. Thus, *excluding* 1, the first few primes are 2, 3, 5, 7, 11, 13, 17, 19, 23, 29, 31, 37….

This is how I see which numbers are primes. First, I write all the positive numbers:

| 1 | 2 | 3 | 4 | 5 | 6 | 7 | 8 | 9 | 10 |
|----|----|----|----|----|----|----|----|-----|-----|
| 11 | 12 | 13 | 14 | 15 | 16 | 17 | 18 | 19 | 20 |
| 21 | 22 | 23 | 24 | 25 | 26 | 27 | 28 | 29 | 30 |
| 31 | 32 | 33 | 34 | 35 | 36 | 37 | 38 | 39 | 40 |
| 41 | 42 | 43 | 44 | 45 | 46 | 47 | 48 | 49 | 50 |
| 51 | 52 | 53 | 54 | 55 | 56 | 57 | 58 | 59 | 60 |
| 61 | 62 | 63 | 64 | 65 | 66 | 67 | 68 | etc | etc |

Then I take away the numbers that are multiples of 2, then those that are multiples of 3, of 4, 5, 6, and so on. The numbers that remain are the prime numbers!

|    | 2  | 3  |    | 5  |    | 7  |    |    |    |
|----|----|----|----|----|----|----|----|----|----|
| 11 |    | 13 |    |    |    | 17 |    | 19 |    |
|    |    | 23 |    |    |    |    |    | 29 |    |
| 31 |    |    |    |    |    | 37 |    |    |    |
| 41 |    | 43 |    |    |    | 47 |    |    |    |
|    |    | 53 |    |    |    |    |    | 59 |    |
| 61 |    |    |    |    |    | 67 |    |    |    |

Interestingly, there is only one even prime number. Does this mean that, with the exception of 2, all the primes are odd? How can I check this without having to calculate one by one? There may be many prime numbers, infinity perhaps, so I cannot check one by one. Surely there must be infinite primes since there are infinite integer numbers, right? If I can add a 1 to any integer number, then no matter how large my last integer would be, I can always make it larger by adding one, which means they have no end. And prime numbers are integers too, thus there must be an infinite number of primes as well. How can I prove this?

Thursday—June 4, 1789

France is in mourning. Today after ten o'clock, the bells in every church in Paris tolled to announce that the Dauphin of France passed away. The young son of King Louis and Queen Marie Antoinette died of tuberculosis. Mother sent us to say a prayer for the soul of the little boy.

After meditating for a while, my mind began to wander. I thought about Pythagoras and wondered why such a brilliant mathematician rejected ratios that are not perfect. Why Pythagoras did not accept numbers like the square root of 2? But then an idea burst in my head. As I gazed up, a shadow from a window formed a right triangle on the wall. The two sides of the triangular shape seemed to be the same size, so I arbitrarily made them each equal to 1. Then I asked, what is the length of the hypotenuse of such right triangle? To answer, I recalled Pythagoras's theorem:

$$x^2 + y^2 = z^2$$

In this case, $x = 1$ and $y = 1$, therefore $z^2 = 2$. This means that I have to find a number which, when multiplied by itself, gives me the number 2. The hypotenuse $z$ is equal to the square root of 2; i.e., $z = \sqrt{2} = 1.414213\ldots$. Thus, $\sqrt{2}$ is an irrational number because it cannot be written as a ratio of integers $p/q$. Why Pythagoras did not see that, too?

Wednesday—June 10, 1789

This morning I announced my desire to be a mathematician. How I regret my childish outburst. How I wish I didn't say it! We were eating and, as usual, my sister was chattering away when, without thinking, I blurted it out. Mother's sweet smile froze on her lips, and then she shook her head but said nothing. Angelique swallowed fast and interjected, "But Sophie, mathematics is not for girls!" and she looked at me as if I was mad.

I was about to explain to her that women can also be mathematicians like Hypatia, but before I could utter a word my father turned to me and, patting my hand tenderly, said, "If Sophie wants to be a mathematician, then she *will be* a mathematician." I wanted to kiss him and hug him as I used to do when I was a little girl. But Mother's austere look kept me frozen in my seat. Surprisingly, Angelique did not comment further; she kept blabbering about her new dress.

After Father left the table my mother admonished me. She thinks it is unbecoming of a young lady to speak such nonsense and chastised me for "filling my head with such outrageous ideas." What is wrong with my wish to become a mathematician? I don't understand why Mother is against my desire to study something I like so much. She claims that mathematics is not part of my education. She forbade me to stay up at night, reading such "an unfeminine subject." Who decided mathematics is unfeminine? Why should mathematics be just for men? I don't accept it!

Still, I wish I didn't say anything at all. But there is no turning back now. Yes, I will become a mathematician somehow! Mother cannot prevent it. There is nothing else in the world that interests me more. I do not want to spend my life in front of the mirror curling and powdering my hair. I only want to study mathematics.

Monday—June 22, 1789

My mother gave me a stern lecture. She discovered that I was reading in bed past midnight, disobeying her orders. But that is not all. She walked into my bedroom when I was writing equations and I didn't have time to hide the notes in my lap desk's drawer. I would have spent the entire night studying, but Mother came and forced me to bed. If I were writing a letter she would not have made such as fuss, but discovering I was working on mathematics made her very angry. She seized my notes, saying she'd burn them; I cried almost all night.

In the morning before breakfast, Mother scolded me again. I don't understand why my mother thinks that mathematics is an unfeminine subject, not proper of a girl. Why does my desire to study mathematics anger her so much? I wish Papa were here. He knows I like numbers; he will stand by me, I am sure of that. After all, I studied arithmetic under his tutelage.

My father is very busy with his work in Versailles and will not return until next week. The assembly with King Louis turned into an ugly affair due to the disputes and arguing of the representatives of the Three Estates. Monsieur Morrell came this afternoon to bring a message from Papa. He told Maman that the arguing became so heated that the deputies of the Third Estate declared themselves the National Assembly, and then they made a pledge not to separate until they give France a constitution.

Millie came to my bedroom before she extinguished the lamps in the hallway. She brought me a cup of hot chocolate and offered to help me finish the handkerchief my mother asked me to stitch. Millie knows that I would rather read a book than do these chores, so I accepted gladly. While Millie embroidered, I read the history of mathematics until I reached the end of the book. I did not notice that Millie left, so it must be late.

It's so easy to get lost in the pages of this book. There are unforgettable stories about the ancient mathematicians, each one pointing me in a different area of study. Reading the history of mathematics makes me realize how much I have yet to learn. For now, I snuff out my candle before Maman finds out I am still awake.

Thursday—June 25, 1789

Millie saved my notes! She found the pages that Maman threw away in the trash bin and hid them for me. I am so relieved! I thought I had to start all over again. Mother is not here this evening. She is at the theater chaperoning Madeleine and her fiancé. This allows me a few hours to myself. I know Maman disapproves and would be terribly angry if she discovers I am back at my studies, but I am very curious about something that Antoine-August mentioned the other day. His tutor is teaching him *polynomials*. He did not say what a polynomial is, but I was intrigued and so I looked it up. I found that a polynomial is defined as any mathematical expression consisting of a sum of a number of terms. What does it mean exactly?

If I have any three terms, such as $x, y, z$, I could write a polynomial as:

$$x + y + z$$

But something is missing, unless I add an equal sign and make it an equation, like $x + y + z = 0$. Now, if I use the same variable $x$ raised to a power like $x$, $x^2$, or $x^n$, then I can define a polynomial as the sum of these terms:

$$x^2 + x^3 + x^4 + \ldots + x^n = 0$$

Is this correct? Oh, how can I be certain? If a polynomial is a mathematical expression involving a series of powers in one or more variables multiplied by coefficients, then I can write a polynomial in one variable with constant coefficients as:

$$a_n x^n + \ldots + a_3 x^3 + a_2 x^2 + a_1 x + a_o = 0$$

The highest power in the polynomial is called its order. And it makes more sense if I write the polynomial as equation, such as:

$$x^3 + 2x^2 + 3x + 5 = 0$$

Then I say that this equation is a polynomial of order 3. Very well; now I just need to learn how to solve it. As Papa always tells me, "Do not give up, *ma petite élève*." Even if I don't have a teacher like Antoine-August, I must try.

Sunday—June 28, 1789

Another protest shook the city with increased violence. Early this afternoon, while reading in Papa's library, I was startled by shooting and chanting. The angry voices jumbled with the sound of firearms, shattered the quietness of the day. Another riot, I thought, and ran to the window. From there I spotted a mob of armed workers and market women, brandishing pikes, and walking westward. The demonstrators shouted slogans full of hate: "Hang the aristocrats!" and "Kill the despot rich; kill them, kill them!"

Angelique was terribly frightened. By the pallor of her face, I could tell Maman was also anxious. She shooed me away from the window and sent us all to her study. We waited until the mayhem

26

subsided, wondering if Papa was caught in the middle of the riot on his way home.

The protests are occurring more and more frequently, and each time they turn more violent. Many people are angry with the king, seeing how he lives in luxury, seemingly oblivious to the suffering of his subjects. Workers in the city are losing their jobs; they resent the unfair taxation and blame Louis XVI for the dwindling economy.

I hope the meeting of the deputies and King Louis in Versailles mitigates the tensions. The newspapers are full of articles written by liberals who promote the formation of a French republic. The liberal reformers want to take away the power of the king and form a government elected by the people. When Papa and his friends discuss these issues my mother just listens and does not say much, but she supports the king and expresses those feelings in private. She does not like the idea of taking away the king's power, as Papa suggests.

Mother is very fond of Louis XVI and Marie Antoinette. She still talks about the royal wedding as if it happened yesterday, but it was in 1769. Mother tells us that Versailles was ablaze with fireworks and feasting, celebrating the nuptials of the young prince and Marie Antoinette, who was fourteen years old at the time. Well, now they're both adults and, having become king and queen of France, they must learn how to rule our nation.

Thursday—July 2, 1789

We were almost trampled by a mob! When we returned home from visiting Madame de Maillard, we found ourselves in the middle of a march on Rue Saint Honoré. The angry protesters were on their way to the Palais Royal. That's the most popular gathering place where leaders of the revolutionary movement preach against the monarchy. The protesters were chanting songs of rage against the king.

Maman, my sisters, and I were trapped in the middle. The faces of the men and women in the crowd were contorted with

27

anger. My mother tried to stay calm, but I noticed how nervous she was by the way she pressed her hands to her heart. We had to wait on the side of the street for the mob to continue their march before we could pass. Angelique kept innocently asking, "Maman, why are those people shouting? Why are they so angry?" but Maman just embraced her protectively. I was scared too at the sight of such rage and held my mother's arm.

When we returned, I escaped into my father's library, the only place I feel secure. The demonstration persisted into the evening. The windows of the study were open because of the summer heat, so I could hear the protesters passing by, inviting people to join them. They were chanting, "*Liberté, Égalité, et Fraternité!*" Some were also intoning slogans of hate I wish I did not hear.

Papa and his colleagues had expected these demonstrations to break out at any time, as many workers are affected by France's failing economy. The causes of the people's turmoil and rebellion include the shortages of bread, the increase in food prices, and the loss of wages. Some leaders are threatening violence if the government does not do something to ease the workers' suffering. But nobody offers any practical solutions.

Yet there's nothing I can do, so I continue solving polynomial equations. The linear polynomial equation $ax + b = 0$ is the easiest to solve since $x = -b/a$. For example, $2x + 3 = 0$ has the solution $x = -3/2$. These are rather trivial; I would like to solve more challenging equations.

First, I need to learn the rules of algebra. But this will have to wait until tomorrow because it is already late. If Mother sees the light of my candle at this hour, she will be furious.

Wednesday—July 8, 1789

Who invented the symbolic notation of algebra? Who was the first mathematician to use one letter of the alphabet to represent the unknown quantity in an equation? In ancient times equations were written using word statements, since algebraic notation was not invented until much later. In the beginning, an equation was stated in words like this example: *A heap, its whole, its seventh, it makes 16.* A "heap" represented the unknown. With modern notation I could write this statement as an equation, using the letter $x$ as the unknown: $x + \frac{1}{7}x = 16$. In this form, any mathematician in the world can interpret it, regardless of what language she speaks, and she would understand that the equation requires that one find the value for $x$ that makes the equation true.

Diophantus of Alexandria was perhaps the first mathematician that made such problems more concise, leading the way for later mathematicians to develop the algebraic notation used today. I prefer to work with symbolic equations, but it is also a good exercise to start with a verbal statement and then rewrite it in symbolic or algebraic form.

The other day, just for fun, I asked Angelique, "Do you know when Papa will be twice as old as I?" My little sister did not know, of course, but I showed her that I could use algebra to answer the question. I can write an equation that includes what I know, and what I don't know, which I call $x$ (the variable). Starting with the fact that Papa is 63 years old and I am 13, after $x$ number of years Papa will be $(63 + x)$ years old, and I will be $(13 + x)$ years old.

I derive a mathematical expression to help me find out when Papa will be twice my age:

$$(63 + x) = 2 (13 + x)$$

and then I solve for $x$ and get,

$$x = 63 - 2(13) = 37$$

This means that in 37 years, Papa will be twice my age; I will be 50 years old ($13 + 37$), and he will be 100 ($63 + 37$). "And you, my dear sister," I said, "will be 47 years old!"

*Très bien.* These are trivial problems with very easy solutions. Now, suppose that I have to solve this equation: $2x^2 + 4 = 20$. In this case, the variable $x$ is raised to the second power, so I have to use a square root to solve it. First, I divide each term by 2 to

29

get $x^2 + 2 = 10$, and then I solve, $x^2 = 8$. Thus, after taking square root, I get $x = 2.82\ldots$.

Wait a minute—shouldn't I get two answers? Since a number $x$ multiplied by itself yields the square of $x$, (or $x^2$), $x$ can be negative or positive. For example $x = 2$ results in $x^2 = 4$, and $x = -2$ also results in $x^2 = 4$. So, either negative or positive values of $x$ will give the same square. This means that a second-degree linear polynomial equation will have two solutions. Should I assume the degree of the equation determines the number of solutions it has? Interesting!

Tuesday—July 14, 1789

W hat a frightful day! Violence and chaos erupted in Paris, again. A huge mob of angry workers stormed the Bastille prison and hundreds of people died. It is awful!

It began this morning. My sisters and I were in Mother's study taking our daily lesson. We heard shooting and at first Maman paid no attention. But soon the racket in the streets was so unnerving that it was impossible to ignore. We ran to the windows and watched hundreds of armed people running, chanting and waving their fists. The multitude moved east towards the prison, chanting so stridently, *"A la Bastille! A la Bastille!"*

After the mob disappeared from sight, we resumed the lesson. I didn't imagine the massacre that was about to happen. But when Papa came back from Versailles in the evening, he told us. A gruesome fight took place when the enraged workers arrived at the state prison. It was supposed to be protest against the increase of taxes. But the workers were defiant and the violent protest ended in massacre. The prison guards could not control the crowd, and many were killed. The Marquis de Launay, the governor of the Bastille, had his throat cut on the steps of the Hôtel de Ville, the town hall, and his head was paraded around the streets of Paris. I was shocked to learn of such horror.

Papa doesn't approve of violence, but he understands why the workers have taken to the streets of Paris demanding social rights. Father sees himself as a moderate revolutionary who owes greater allegiance to his nation than to his monarch, as he supports the ideals for social equality. My mother is unswerving in her loyalty to King Louis; she doesn't understand why he is blamed for what is happening. Of course she is sympathetic to the plight of the poor, but she doesn't believe that fighting is the way to solve the social problems in France. I agree.

Saturday—July 18, 1789

A new era for France begins. The newly formed Constituent National Assembly began to rule. Its members are drafting a new constitution. The king came from Versailles with some of his retinue, but without the queen. Louis XVI arrived to the Hôtel de Ville, where the new mayor of Paris, Monsieur Jean Sylvain Bailly, received him, surrounded by a large crowd of citizens.

Mother wanted to go, but my father did not think it prudent to be near the city hall. He was afraid another protest could erupt since many people are unhappy with His Majesty. The meeting, however, went on smoothly. At one point, Papa related, the king placed the blue and red ribbon on his hat as a gesture of support for his subjects.

Monday—July 20, 1789

When Millie came this evening, she seemed curious about the numbers on my paper. I shared my study of algebra, a branch of mathematics that uses mathematical statements to describe relationships between things that vary over time. To make it simpler, I told Millie that when I use a mathematical statement to describe a relationship, I use letters to represent the quantity that varies, since it is not a fixed amount. These letters and symbols are referred to as variables. For example, it can describe the relationship between the supply of bread and its price.

It was hard to explain algebra to a young woman who doesn't read or write. I used other simple examples that Millie could understand. I told her that ancient mathematicians did not write equations like the ones she sees in my notes. The notation I use was not invented until later.

For example, if she'd ask how old I am, I could reply, "My age, plus 16, is equal to three times my age minus 10." Using algebra, she could figure out my age by writing my words as an algebraic equation, using the letter $x$ as my age, which is the unknown number:

$$x + 16 = 3 \cdot x - 10$$

The first expression stands for "my age in years plus 16," written as $(x + 16)$. This is equal to the second expression for "three times my age, minus 10," written as $(3 \cdot x - 10)$. The equation has a variable $x$, the unknown in this case, so the solution to the equation is the number that makes the equation true. To solve the equation I put all the terms that contain $x$ on one side, and all the other terms on the opposite side:

$$x - 3 \cdot x = -10 - 16$$
$$2 \cdot x = 26$$

The answer is: $x = 26 \div 2$, or $x = 13$ (indeed, this is my age). This means that replacing each occurrence of $x$ with 13 makes the original equation true, since

$$x + 16 = 3 \cdot x - 10$$
$$13 + 16 = 3 \cdot 13 - 10$$
$$29 = 39 - 10$$
$$29 = 29$$

*Voila!* Solution is verified!

I can make up any algebraic equation related to some unknown quantity. They are fun and easy. Millie was amused but did not want to see more examples. She had chores to do before going to bed. I was left alone to solve more equations. Of course, I didn't tell her that a mathematical statement is not just for variations of real objects. For example, an algebraic equation such as $x^2 + 1 = 0$ does not have to refer to anything in particular. That is why learning algebra requires abstract thinking.

Now I better put away my notes and get ready for bed.


Wednesday—July 22, 1789


I've been reading about the Greek mathematician Diophantus of Alexandria. The author of one history book states that nobody knows exactly when he was born or when he died. The entry in the encyclopedia simply indicates that Diophantus lived about 250 A.D. Then I consulted the *Greek Anthology*, which is a collection of epigrams, and found among its mathematical problems a riddle about the life of Diophantus:

*God granted him to be a boy for the sixth part of his life, and adding a twelfth part to this, He clothed his cheeks with down; He lit him the light of wedlock after a seventh part, and five years after his marriage He granted him a son. Alas! late-born wretched child; after attaining the measure of half his Father's life, chill Fate took him. After consoling his grief by this science of numbers for four years he ended his life.*

What I infer from this statement is that the son of Diophantus was born after $1/6 + 1/12 + 1/7$ of his life plus 5 years. The son died 4 years before Diophantus, and lived half as long. To solve the puzzle mathematically, I write everything in terms of the age of Diophantus, or number of years he lived, and, since it is unknown, I call it $x$. Now, according to the riddle, his son was born in the year equal to $(1/6 + 1/12 + 1/7)$ times $x$ plus 5. In equation form this can be written as

$$\left(\frac{1}{6} + \frac{1}{12} + \frac{1}{7}\right)x + 5$$

33

I also know that the son died 4 years before Diophantus died, so I use the relation $(x - 4)$. I then subtract the year of birth from the year of death, and get the son's lifespan, which is half of his father's:

$$(x - 4) - \left[\left(\frac{1}{6} + \frac{1}{12} + \frac{1}{7}\right)x + 5\right] = \frac{x}{2}$$

This equation reduces to $\quad \frac{9}{84}x = 9$

Solving for $x$ the age of Diophantus, I find: $x = 84$

From this I deduce Diophantus married at age 33, had a son when he was 37, and died when he was 84. *Voila!*

Friday—July 24, 1789

Monsieur du Chevalier is moving to Sweden! He did not come to give me a piano lesson today. Instead, he sent his valet to notify my parents M. du Chevalier will no longer teach me because he is leaving France. Papa was not surprised. Members of the First Estate are angry with King Louis for taking away their privileges. Others are afraid of the revolutionary movement, and to avoid persecution, many aristocrats are emigrating.

Since my lesson was cancelled, Maman gave me permission to go to the bookshop down the street and buy a book. I love to go there; the smell of old books reminds me of the times I spent with Papa in his library when I was a little girl. He used to show me his childhood books and taught me arithmetic from them.

I was nervous to go alone to the bookshop because Monsieur Baillargeon, the proprietary, intimidates me. He is an old, gaunt man with a permanent scowl on his forehead that makes him look perpetually angry. But I mustered the courage to ask for a book on mathematics. M. Baillargeon looked at me strangely, as if he did not understand me, and then whispered, "Perhaps mademoiselle wishes to buy a nice romance novel." I thanked him politely and repeated that I wanted a book *with* mathematics. After a few moments' hesitation, M. Baillargeon smiled condescendingly, but scurried to the back of the shop. He returned with an old dusty book and handed it to me.

I eagerly flipped through the yellowed pages. The book was old and fragile, but none of the pages seemed to be missing. With a few illustrations and many equations, the book looked interesting and cost only twenty-three *sous*. M. Baillargeon kept looking at me inquisitively, so I paid him and left in a hurry to get away from his prying gaze.

The book is a treasure. It's a translation of the *Arithmetica,* a famous work written by Diophantus hundreds of years ago. Diophantus, from the Greek city Alexandria, was perhaps the first to develop algebra. Ever since I read the story of Hypatia, I was curious about him. Hypatia wrote about *Arithmetica* and taught it to her pupils. That is why I must study it.

We are going for summer vacation tomorrow morning. My mother is anxious to leave the foul city air and refresh our spirits in the country. We have a home in Lisieux, about 150 kilometers from Paris, where we spend one or two months every year. We normally depart in late June; this time, however, because of my father's business and his affairs with the National Assembly, it was delayed. I look forward to this holiday in the countryside. Although the voyage is long and tiring, our home in Lisieux is a welcome respite from the summer heat in Paris.

Sunday—July 26, 1789
Lisieux, France

After an exhausting trip, we arrived in Lisieux. We left Paris yesterday morning and the long journey took a day and a half. At times, it felt hot and crowded in the rickety stagecoach, but I immersed myself in a book and tried to ignore my sister's incessant chatter.

We stopped at a tiny village to eat lunch and exchange the horses. After a short rest we resumed the trip and rode for another six hours. We arrived in Rouen at sunset. The inn where we spent the night was an old château, dark and ugly. Before dinner, Angelique and I visited the ancient chapel nearby and explored its decaying cemetery. While reading the inscriptions in the tombs, trying to imagine the lives of the people who were buried there, we were startled by an odd looking man. Toothless and with an ugly scar on his cheek, the old man resembled a character in a horror tale. Limping, the man approached us smiling, as if happy to see us.

After chatting nonsense, the old man pointed at a fairly new grave. He said the daughter of the Marquis who owns the inn was buried there. He added that the young girl died of fright one night when she met the ghost of the castle. He was probably making up the story, for he kept smiling mischievously. But Angelique was so terrified afterwards that she had to sleep in Maman's bed. In the morning, after an unappetizing breakfast, we continued our trip.

Lisieux is a quaint town in the *Pays d'Auge* region in northwest France. Our house is located a few miles from the village, hidden among gentle hills, valleys, and forests. Lisieux landscape is beautiful, dotted with old manor homes, some châteaux, and half-timbered farms. The region is enchanting, and the smell of cider mixes with the aroma of cheese. As we approached the *Pays d'Auge*, the monotony of the long ride from Paris was broken by the sight of thatched roofs, winding brooks, charming chapels, and peaceful hamlets as colorful as those seen in old paintings.

We arrived in time for a delicious supper. Margarite, the lady who keeps the house, and her daughters had the table ready with cold meats, green salad, and refreshing lemonade. We spent the afternoon unpacking and reacquainting ourselves with the village where we attended evening Mass.

It is late already, and Maman and my sisters have gone to sleep. The house is quiet, and I sit here, listening to the sounds of the country, watching the flame of my candle being swayed by the gentle breeze. A fresh bed with a down feather mattress awaits me to restore my tired body. Margarite put sachets of flowers in my pillow to attract happy dreams, she claims. Oh yes, I must rest and dream of what I'll discover in my books next.

I am studying the *Arithmetica* of Diophantus. It deals with the solution of algebraic equations and the theory of numbers. *Arithmetica* is a collection of problems with numerical solutions of indeterminate and determinate equations (those with unique solutions). Although parts of the book were lost, probably even before Hypatia's time, the parts that remained contain the solutions of many problems concerning linear and quadratic equations, but it considers only positive rational solutions to these problems.

Diophantus studied three types of quadratic equations: $ax^2 + bx = c$, $ax^2 = bx + c$, and $ax^2 + c = bx$. The reason why there were these three cases is because Diophantus did not know the number zero. He also avoided negative coefficients by considering the coefficients $a$, $b$, and $c$ to be positive in each of the equations.

I think I know why Diophantus did not include negative numbers. If one thinks of numbers for counting physical objects, like apples or houses, then "minus 3 apples" or "−7 houses" make no sense. However, if one thinks of numbers as ideas, as mathematical objects, then one can readily accept numbers having negative values. For example, if I think of numbers as points along an infinite line, then I must include negative numbers. With zero as the middle point on the line, I can see numbers along the right side of the line (positives) and numbers on the left side (negatives). Otherwise the line would end at zero, which makes no sense, unless zero was positioned at infinity!

But, if the solution to the simple linear equation $ax + b = 0$ was known since ancient times to be $x = -b/a$, then why Diophantus did not include the negative numbers? The easiest way to interpret this equation is by writing $a = 1$ so that $x + b = 0$; the value of the variable $x$ is simply the negative value of $b$ ($x = -b$). For example, in the equation $x + 5 = 0$, $x$ has to be equal to −5 in order to make the right hand of the equation zero. There it is, unmistakably stated; we need negative numbers to solve this equation.

Friday—August 14, 1789
Lisieux, France

Now I know how to solve polynomials. I started by learning that polynomial equations consist of terms involving an unknown number $x$ raised to a positive whole power or exponent and multiplied by constants $a$ and $b$. Such equations are classified by the largest exponent of the unknown $x$.

First-degree equations are those like $ax + b = 0$. Second-degree equations are like $ax^2 + bx + c = 0$. Third-degree equations are like $ax^3 + bx^2 + cx + d = 0$, and so on. Either of the constants can be equal to zero, in which case the number of terms is reduced. The important thing to look for is the largest exponent of the unknown number.

Solutions to this kind of polynomial equation will always be algebraic numbers. For example, linear equations that have the general form $ax - b = 0$, where $a$ and $b$ are two known numbers, have solution $x = b/a$. This would be a fraction, zero, or an integer, depending on the values of $a$ and $b$. This implies that all linear equations have solutions that are rational numbers.

I've solved many first-degree equations. They're very easy because the equations have a single variable, say $x$, raised to the first power. For example, $ax + b = 0$, in which $x$ is the same as $x^1$. In other words, $x$ means "$x$ raised to the first power."

I also solved equations of the form $ax^2 - b = 0$. The solutions are of the form $x^2 = b/a$. Again, depending on the values of $a$ and $b$, I get a solution that is an integer, a fraction, or a radical. I solved many equations, every one of which I could find in my books, and then made up some myself.

Now, let's consider a circle of diameter $d$ and perimeter $p$. If I divide the perimeter by its diameter, the result always gives the same number, and this number is $\pi$, no matter how big or small the circle is. In other words, $p/d = \pi$. I know that $\pi$ is *not* a rational number. Does it mean that $\pi$ is not an algebraic number?

Our summer vacation ended abruptly. We returned to Paris yesterday after the riots started in the province. My mother became very nervous when the signs of unrest in the village threatened the peace around us. The revolutionary movement is spreading very quickly, causing rioting and pillaging in places that used to be so quiet. Incited by a few, the farmers who worked for the rich landowners all their lives are taking back the land, using all means possible, including pillaging and violence. We witnessed a family fleeing their estate, terrified by the angry mob of peasants raiding châteaux and burning land titles.

Paris is also in a state of chaos. Last night, the archbishop's house was vandalized. We heard windows breaking and the mob's shouting woke us up around three in the morning. Millie saw later in the daylight that the property was badly damaged. The workers were furious with the rich aristocrat abbots who live in luxury while they live in misery. The people resorted to violence to express their frustration.

Papa brought home a copy of *The Declaration of the Rights of Man and of the Citizen* just issued by the National Assembly. The declaration states that all men are born and remain free and equal in rights. It also implies that all citizens of France will share the political power. However, it only mentions men. Why are women not included? Don't women have the same rights? I should have asked Papa, but I didn't have a chance. My sister Madeleine came in with her fiancée Monsieur Lherbette and everybody started to congratulate them on their engagement. They will marry sometime this year.

Antoine-August came to visit with his parents. He asked me to help him solve a quadratic equation of the form $ax^2 + bx + c = 0$, where $a$ is a nonzero number and $x$ is the unknown variable. Imagine, he asked me for help! I told him one way to solve the equation; by applying the formula: $x = \dfrac{-b \pm \sqrt{b^2 - 4ac}}{2a}$

He had to solve very simple equations like this: $3x^2 - 4x = 2$. The first thing I did was rewrite it in the form of the quadratic equation:

$$3x^2 - 4x - 2 = 0$$

where $a = 3$, $b = -4$, $c = -2$. Then I showed him the solution:

$$x = \frac{-b \pm \sqrt{b^2 - 4ac}}{2a} = \frac{-(-4) \pm \sqrt{(-4)^2 - 4(3)(-2)}}{2(3)} = \frac{4 \pm \sqrt{40}}{6}$$

I should say the *two* solutions because $x$ has two different values, $x = \dfrac{4 + \sqrt{40}}{6}$ and $x = \dfrac{4 - \sqrt{40}}{6}$. Either value of $x$ solves the equation.

Quadratic equations are easy; I am sure Antoine-August could have solved his without my help, but I am glad he asked.

Saturday—September 5, 1789

Mathematical analysis is amazing! It requires the application of specific rules and logical thinking. I discovered that sometimes there are multiple ways to arrive at the final answer. Take for example polynomial equations of second degree. These are quadratic equations of the form $ax^2 + bx + c = 0$, where $ax^2$ is the quadratic term, $bx$ is the linear term, and $c$ is the constant term. I know three methods for solving them: by factoring, by completing the square, and, of course, by applying the quadratic formula. In all three cases it is easier to reduce the polynomial to its monic form with the coefficient of $x^2$ equal to 1, such that $x^2 + px + q = 0$, where $p$ and $q$ are real numbers.

For example, the following equation is easy to solve by factoring: $x^2 - x - 2 = 0$,

$$x^2 - x - 2 = (x + 1)(x - 2) = 0$$

where I can equate the binomial expressions to zero and solve them individually:

$$x + 1 = 0; \qquad x = -1$$
$$x - 2 = 0; \qquad x = 2$$

Of course, I could solve this equation with the quadratic formula, but it is much faster and easier to do it by factorization.

For quadratic equations that lack the constant term, such as $x^2 + px = 0$, I can solve them with the method I call "completing the square." The idea is to make the binomial expression a perfect square by adding a quadratic supplement $(p/2)^2$ to complete the square, as follows:

$$x^2 + px + (p/2)^2 = (x + p/2)^2$$

For example, to solve the equation $x^2 - 3x = 4$, I add $(3/2)^2$ to both sides of the original equation to get:
$$x^2 - 3x + (3/2)^2 = x^2 - 3x + 9/4 = (x - 3/2)^2$$

$$x = 3/2 \pm 5/2$$

So, the two roots or solutions of the equation are

$$x_1 = 4 \text{ and } x_2 = -1$$

There are quadratic equations that do not have the linear term, and in this form the equations are called "pure quadratic," such as $2x^2 - 9 = x^2$, which I can solve by reducing it directly: $x^2 - 9 = 0$, so that its two roots are $x = \pm 3$. In this case there is no need to use the quadratic formula.

Antoine-August asked which method is better to solve quadratic equations. I told him that it depends on the equation. All the methods are correct, but a mathematician decides which one to use based on preference, or whether one wishes to do it in the simplest or fastest way possible.

He also mentioned cubic polynomial equations and asked if I knew how to solve them. I replied, rather arrogantly, that if I could

41

solve quadratics then I should find solutions for cubics as well. I better learn how before Antoine-August discovers I don't know. Otherwise he'll never ask me for help again.

$C$ubic equations are more difficult. Starting with the polynomial $a_n x^n + \ldots\ldots + a_3 x^3 + a_2 x^2 + a_1 x + a_o$, if the exponent is $n = 3$, the equation becomes $ax^3 + bx^2 + cx + d = 0$ and is called "cubic." These equations were fully solved hundreds of years ago using geometrical methods. However, solving the cubic equation using pure algebra was a great challenge to mathematicians. In fact, it proved so tough that the Italian mathematician Luca Paciola wrote in his book *Summa de Arithmetica* that "the general cubic equation is unsolvable."

The solution was obtained in the sixteenth century. Before that, mathematicians could only solve special cases of the cubic equation. For example, Scipione del Ferro could solve the depressed cubic, $ax^3 + cx + d = 0$. He kept it a secret because, at that time in Italy, new discoveries were used as weapons against opponents in mathematical contests. Just before his death, del Ferro handed the solution to his student, Antonio Fior. Then Niccolo Fontana, better known as Tartaglia, challenged Fior. To respond to Tartaglia's challenge, Fior retaliated with thirty problems on depressed cubics. Eventually, after much effort, Tartaglia solved the problems.

I didn't realize how important it could be for a mathematician to fight for the honor of being the first one to solve a difficult problem. Now I am more determined to learn how to solve these equations and try to be the best. Interestingly, Niccolo Fontana was known by the nickname Tartaglia, which means in Italian "one who stutters," because of his speech impairment. Some said he stuttered due to a wound he suffered in a duel.

Now I can also answer the question, is $\pi$ an algebraic number? I can categorically say no. I reason as follows: by definition $\pi$ is equal to the ratio of the circumference of a circle to its diameter, no matter how one measures. If $\pi$ were an algebraic number, then it would be a solution to some polynomial that has integer coefficients. It is not. Thus, if $\pi$ is not an algebraic number, then it must be a number that goes beyond, that transcends the algebraic numbers. I know that $\pi$ is not a whole number or a perfect fraction, and that the decimal part does not show a repeating pattern.

Oh, I am digressing. I must learn how to solve cubic equations before Antoine asks for help again.

Friday—September 18, 1789

$S$till struggling with cubics! At first I thought cubic equations of the form $x^3 + ax^2 + ax^1 + a^0 = 0$ were easy to solve. But soon it became clear they are more challenging. At least the procedure is not straightforward as that for solving quadratic equations! About 200 years ago, an Italian mathematician named Girolamo Cardano published the solution to the cubic equation. I wish I knew it because I've tried solving a simple cubic and failed. It is so frustrating!

Two nights ago, I attempted to solve a simpler cubic like this: $x^3 + mx = n$. It looked easy enough but, after trying for several hours, I couldn't find a solution. What am I missing? I will not sleep well if I do not solve it. I will attempt a different approach.

First I notice that $(a - b)^3 + 3\,ab(a - b) = a^3 - b^3$

Then, if $a$ and $b$ satisfy $3\,ab = m$, and $a^3 - b^3 = n$, then $a - b$ is a solution of $x^3 + mx = n$.

Now $b = m/3a$, so $a^3 - m^3/27a^3 = n$, which I can also write as

$$a^6 - na^3 - m^3/27 = 0.$$

This is like a quadratic equation in $a^3$,

$$(a^3)^2 - n(a^3) - (m^3)/27 = 0.$$

43

So, I can solve for $a^3$ using the formula for a quadratic equation.

*Très bien*. Then I can find $a$ by taking cube roots. I can find $b$ in the same way, or by using $b = m/3a$. Therefore, $x = a - b$ is the solution to the cubic equation above.

Am I correct? Well, I can substitute $x$ into the original cubic equation and see if it's true. I may ask Antoine-August to show my analysis to his tutor and see if I'm doing this correctly.

Friday—September 25, 1789

Autumn is slowly making an entrance. After lunch my sisters, my mother, and I walked to the *Jardin des Tuileries*. As we made our way to the gardens, Madeleine's fiancé, Monsieur Lherbette, joined us. It was good that he came, as it made the stroll more enjoyable. The streets are swarming with indigent people, barefoot children, and haggard women begging for money. I noticed more beggars angrily shaking their fists at the carriages of the wealthy citizens. But the sight of poor children is heartbreaking. I gave the few coins in my purse to a pretty little girl dressed in rags. The poor child, she didn't say a word but eagerly took the coins and ran.

When we were near the Louvre Palace, M. Lherbette told me that the French academies have residence there, including the Academy of Sciences. There is where the best mathematicians in the world work on their research. I was thrilled to know I was walking so near the great men of science. M. Lherbette added that every week the Academy of Sciences holds public lectures where the scholars expound their work. Wouldn't it be great if one day I could attend the lectures? As we went past the massive front doors of the old edifice, I saw gentlemen with stern faces coming and going, but I could not discern who they were.

One day, I know, I will meet a mathematician, and perhaps one day he'll invite me to attend the lectures at the Academy of Sciences.

To be a mathematician one must prove theorems and discover new ones. I want to be a mathematician. To do it I must use my own reasoning to prove concepts that might not be obvious. The ancient Greek scholars developed a mathematical method to prove theorems that they called *reductio ad absurdum,* which means "reduction to absurdity." With this method, one assumes the opposite of what one desires to prove. Then one can show that this assumption leads to a contradiction. Or, if the opposite of what one wishes to prove is false, the statement to be proven must be true.

I could learn this method by verifying a theorem. What better example than proving the irrationality of numbers. In general, to prove that a certain number is irrational, I could start by assuming that the number is represented by some ratio of whole numbers, say $p/q$. If I can find a fraction $p/q$ equal to the number, then the number is rational; however, if $p/q$ is not equal to the irrational number, then I can conclude that no fraction exists equal to the number, and therefore it is an irrational number. It seems simple enough; I am sure I can prove that the square root of 2 is an irrational number.

For now I will entertain myself with a game of numbers. I start with perfect squares. By definition, a perfect square is a number that can be written as a whole number times itself. For example, 81 is 9×9, and 25 is 5×5. Thus, 1, 4, 9, 16, 25 are perfect squares. The Greeks proved that any square root of a whole number that is not a perfect square is incommensurable. Therefore, the square roots of 3, 5, 6, 7 and 8 are incommensurable numbers. I also find that the sum of the first $n$ odd numbers is a perfect square, that is $n^2 = 1 + 3 + 5 + 7 \ldots + (2n - 1)$. Curiously, I find that every perfect square ends in a 0, 1, 4, 5, 6, or 9.

Oh, it must be very late; the clock in the drawing room chimed the eleventh hour long ago. I must sleep or tomorrow I won't get up on time for morning prayers. Maman would be terribly upset with me.

The court has moved to Paris. It was not a voluntary move on the part of the king—he was *forced* by the people. It began with the assault of the royal palace in Versailles two days ago. A mob of angry working-class women marched to Versailles demanding to be heard by King Louis. They went to complain that the poor people in Paris can't afford to buy bread.

The demonstrators wanted to talk with His Majesty, to let him know they need help to resolve the economic problems. Armed with pitchforks, the women walked for six hours to Versailles. Upon arrival, the enraged women stormed the château, ran up the queen's staircase, and broke into the royal chambers. Queen Marie Antoinette had to run to join the king and their children. How terribly scared she must have been, seeing hundreds of angry people brandishing firearms and pikes, shouting insults and running through the palace.

We know all this because Papa witnessed the assault. He said that there were mothers driven to despair because they could not feed their children. After storming the palace, a delegation of women gained an audience with the monarch; they also addressed the National Assembly. The protesters complained that the wealthy citizens are hoarding grain; some claim that is why there isn't enough bread for the poor.

My mother blames the leaders of the revolutionary movement for this revolt. She is a generous woman with a tender heart. In fact, she works with the *Charité Maternelle*, the Maternal Charity founded by Mme Fougeret, to help poor women care for their babies. So, when she heard about the women marching to Versailles demanding bread, she understood their plight. However, she thinks someone incited the women to violence. We do not know how all this started.

King Louis promised to produce bread for everybody, and he allowed the women to escort him, his family, and his court back to Paris. This way, the king will be within reach of the people. The women assured His Majesty that in Paris he would find faithful advisors who can tell him how things stand with his subjects, enabling him to act accordingly. Louis XVI and his family now reside in the Tuileries Palace nearby. The National Assembly also moved to a hall building close to the Tuileries. This is good news for us. Papa does not have to ride to Versailles anymore. I'm glad because he'll be back to play chess with me. I can't wait to tell him about my studies.

$T$o prove the irrationality of the square root of 2, I use the *reductio ad absurdum* method. I start the proof by contradiction by assuming that the square root of 2 is rational, and then I can express it as a ratio of two numbers:

$$\sqrt{2} = \frac{p}{q}$$

$p$ and $q$ are the smallest positive integers that satisfy the equation.

Now $p$ and $q$ cannot both be even because, if they were, I could divide each by 2 and have a smaller $p$ and $q$ to work with.

From this I can get, by squaring both sides:

$$2 = \frac{p^2}{q^2}$$

Then by multiplying by $q^2$

$$2\, q^2 = p^2$$

From this I can deduce that $p^2$ is even.

If $p^2$ is even, then $q$ must be odd since $p$ and $q$ cannot both be even. Since $p^2$ is even, then $p$ must be even. If $p$ is even, then $p^2$ must be divisible by 4.

Thus, $\dfrac{p^2}{2}$ is an even integer.

47

However, since $2\,q^2 = p^2$

I should write, $\dfrac{p^2}{2} = q^2$

Thus, $q^2$ must be even, and therefore it follows that $q$ is even. But I had previously determined that $q$ must be odd. And since $q$ cannot be both even and odd, I conclude that no such integer exists, which invalidates my original premise that the square root of 2 is rational. In other words, $\sqrt{2}$ has to be irrational. That's it; I proved it!

Saturday—October 24, 1789

$H$ow did Archimedes confirm that the crown was not made of pure gold? According to the story, Archimedes discovered a method to prove it when he observed that the amount of water that overflowed the bathtub was proportional to the weight of his body submerged in the water. But how? For sure, Archimedes must have used algebra. If I had to do it myself, this is what I would do:

I start by assuming that I know the weight of the crown and the volume of water displaced by it, quantities which I call $m$ and $V$ respectively. If, as the king suspected, the crown was made of gold and silver, the crown weighed an amount equal to the sum of the weight of the two metals, say $m = m_1 + m_2$. This gives one equation and two unknowns. Thus, I also must know the amount of water displaced by a certain amount of gold and a certain amount of silver. I could say that the total volume of water displaced by the crown is equal to the contributions from the volumes displaced by each one of the two metals.

I need to know that the specific volume of a body is equal to its volume divided by its mass, $v = V/m$, so assume for gold $V_1 = v_1\,m_1$ and for silver $V_2 = v_2\,m_2$. Thus, I can write another equation relating the volumes of gold and silver making up the total volume of the crown, $V = V_1 + V_2 = v_1\,m_1 + v_2\,m_2$. Therefore:

$$m = m_1 + m_2 \tag{1}$$
$$V = v_1\,m_1 + v_2\,m_2 \tag{2}$$

48

Now I have a set of two equations and two unknowns, $m_1$ and $m_2$. The simplest way is to solve for $m_1$ from equation (1), and substitute this into equation (2) as follows:

From (1)  $m_1 = m - m_2$  (3)

Substitute (3) into (2)

$$V = v_1 m_1 + v_2 m_2 = v_1 (m - m_2) + v_2 m_2$$

$$V = v_1 m - v_1 m_2 + v_2 m_2$$

or

$$V = v_1 m + m_2 (v_2 - v_1)$$

so,

$$m_2 = (v_1 m - V)/(v_2 - v_1)$$  (4)

Since $m$, $V$, $v_2$, and $v_1$ are known quantities, now I can calculate the weight of silver, $m_2$. And substituting (4) into (3), I obtain $m_1 = m - m_2$, the weight of gold. Obviously, if the crown were made of pure gold the value of $m_2$ would be zero.

Is this how Archimedes proved that the crown was made of gold and silver? Since he was a mathematician, he must have used similar analysis.

Tuesday—November 17, 1789

I have not written in a while because Mother confiscated my paper and pens. She was angry with me after she found me studying all night. I couldn't believe it myself, but the time goes by so fast when I am solving problems. Maman gave me a harsh lecture, again. This morning, Papa talked with her and convinced her that my studies are harmless, as long as I do not stay up for many hours. Reluctantly, Maman agreed to give me back my pens. But I had to promise her I would snuff my candle and go to sleep no later than eleven o'clock.

I must master the theory of numbers. Numbers are the foundation upon which mathematics rests. And the nature of numbers is established by rigorous rules, because not all the numbers are the same. For a merchant, numbers are tools to perform the basic arithmetic computations like adding and subtracting. Most people rely on natural numbers such as the integers 1, 2, 3, ... that is, the ordinary numbers that one uses for counting, adding, and subtracting objects. Of course, the negative numbers and zero can also be considered ordinary numbers, but most people would not think much about negative numbers since they do not use them for everyday computations.

The rational numbers, which are ratios of integers such as 1/2, 5/2, 13/27, are the numbers that the Pythagorean mathematicians believed to rule the universe. This type of number has special rules, such as the one that establishes that a rational number $a/b$ is valid only if $b$ is not equal to zero. In other words, division by zero is never allowed. There are mathematical theorems for rational numbers that prescribe this rule.

One theorem states: "The product of any rational number $c$ and zero is equal to zero, or, $0 \cdot c = 0$." Another theorem states: "$a/b = c$ if and only if $a = b \cdot c$." This theorem says that if the rational number $a/b$ is equal to another number $c$, then it must be true that the numerator of $a/b$ equals the denominator multiplied by $c$. For example: $12/3 = 4$ if and only if $12 = 3 \times 4$. Now, if I write $12/0$, this would imply that there is a number $c$ equal to $12/0$, and by the second theorem I would have to write $12 = 0 \cdot c$. However, the first theorem tells me that the product of any rational number $c$ and zero is equal to zero. How can I have $12 = 0$? This is absurd. Therefore, I conclude that no rational number can have zero as its denominator. This is how the ancient mathematicians established the rigorous rules by which mathematics is built.

I performed endless calculations, starting with ratios such as $5/2 = 2.5$ and $3/4 = 0.75$, but then I discovered that relatively simple rational numbers such as $5/3$ and $7/3$ have infinite repeating decimals, since $5/3 = 1.666666....$ Then I came across other types of rational numbers such as $2/7 = 0.285714285714....$, a number whose integers after the decimal point repeat over and over indefinitely.

It wasn't just my curiosity about how many digits I could calculate. I also wanted to understand the difference between rational and irrational numbers. There is a theorem that states: "any irrational number is represented by an infinite decimal where the string of digits neither terminates nor has a repeating pattern." I used this fact to conclude that the number 2/7 is not irrational since its decimal has a repeating pattern. However, the square root of 2 and $\pi$ are irrational numbers since $\sqrt{2}$ = 1.41421356..., and $\pi$ = 3.1415....

When I do mathematics I lose track of time; I transport myself to another world where nothing else matters. I must be careful, or I could stay up all night and would lose my mother's trust. I promised her to go to bed early.

Saturday—November 21, 1789

It is so cold already. Yesterday, a poor woman froze to death at the steps of the church. How ironic that the archbishop lives in a luxurious palace just a few meters away from where the wretched woman died of hunger and cold. Couldn't he have done something for her? People should not die of hunger and cold at the church steps. It is inhuman.

Papa believes that the economy and the social conditions of the poor will improve. On the 2nd of November the delegates of the National Assembly voted to nationalize church property. This was the Assembly's solution to the increasing economic problems France now faces. The delegates did not want to levy new taxes; instead, they proposed to take over the church lands. The Assembly plans to sell the lands and issue bank notes that will become the new French currency. It's just fair to me that the wealth of the Church is used to help the poor people.

Saturday—November 28, 1789

I'm in trouble! Mother is so angry at me that she took away my candles. A few days ago she found me awake, solving equations, unaware that it was past midnight. Maman was enraged that I disobeyed her, and as punishment she now sends me to bed in darkness. I am only allowed to read in the library during the day. That's fine. I'll use that time to learn more mathematics.

I am writing by the light of an old worn candle I found in the kitchen. I'll go to sleep right after I finish this note. It would be much easier if Maman were more understanding. I would not have to sneak out in the middle of the night looking for candles. And now that winter is approaching, my bedroom gets very cold at night. Oh, Mother, would you ever accept my desire to learn mathematics? Taking away the heating in my room or the candles will not diminish it.

<br>

Saturday—December 5, 1789

Paris is quiet and peaceful tonight, as if the violent events of the previous months never happened. The cold of winter forces people to stay home, so the political instigators have smaller audiences. December is also the time of year when people are in a cheery mood, and families focus more on the Christmas festivities. Whatever the reason for this tranquility, I am glad that the fighting and hostilities we witnessed in the summer have vanished. Paris is again the beautiful city I love. Even the glow from the lampposts seems gentler and warm.

I like this time of year because Paris becomes an enchanted city, lit by thousands of candles, and there is merriment everywhere. I hope the riots of the summer don't happen again.

<br>

Wednesday—December 9, 1789

Papa came to see me, concerned because I was absent at the dinner table. Yes, I was upset because my mother gave me a lecture. She chastised me for spending too much time reading and doing all this *crazy* mathematics, as she calls my studies. Her words hurt me, and that's why I did not feel like eating. I also did not want to run into Madame Charbonneau, who came to visit my mother. She is a silly old woman who always finds reasons to criticize me, especially if she sees me reading, or playing chess. She remarks about my appearance, saying that I don't look feminine, just because I do not put rouge on my cheeks.

I am so glad Papa came to see me. His conversation cheered me up, as always. I showed him my notes, and he was astounded. My father kept looking at the equations I've written and couldn't believe I had done all that analysis by myself. He was astounded when I showed him the solution to the riddle he gave me in April. I solved it weeks ago, after I had mastered algebra. Here I rewrite the riddle and include my solution:

*A long time ago, in a far away land, a sage woman raised three boys to be honest and smart. When her husband died, he bequeathed her one milking cow, and thirty-five cows more to be divided among their three sons. According to the wishes of the father, the oldest was to receive half the number of cows, the second would get a third of the total, and the youngest would receive a ninth of all.*

*The brothers began to quarrel amongst themselves, as they did not know how to divide the herd. They attempted different ways to divide it, and the more they tried, the angrier they became. Every time one of the brothers proposed a way to divide the animals, the other two would shout that it was unfair.*

*None of the suggested divisions was acceptable to all. When the mother saw them squabbling, she asked why. The oldest son replied: "Half of thirty-five is seventeen and one-half, yet if the third and the ninth of thirty-five are not exact quantities either, how could we divide the thirty-five cows amongst the three?" The mother smiled, saying the division was simple. She promised to give each their share of the herd and assured her sons it would be fair.*

53

*She started by adding her own cow to the herd of 35. The sons were stunned and thought she was crazy to give away her only cow. But she assured them there was no need to worry, as she would benefit the most.*

*The woman began: "As you see, my sons, now there are 36 cows in the herd." Turning to the oldest, she stated that he'd get one-half of the 36, which is 18 cows. "Now you cannot argue, as originally you would have gotten 17 and one-half. You gain with my method, right?"*

*Turning to the second, the mother continued: "You, my middle son, will get a third of 36, which is 12 cows. You would have gotten a third of 35 which is 11 and a little more of another, right? You cannot argue either, as you also gain."*

*And to the youngest son the woman said: "According to your father's wishes, you were to get a ninth of 35, which is 3 cows and a fraction of another. However, I will give you a ninth of 36, which is 4 cows. You also gain and should be satisfied with your share."*

*The wise woman ended with these words: "With this division that favors all, 18 cows are for the oldest, 12 for the middle, and 4 for the youngest son. This gives a total of 34 cows (18 + 12 + 4). Of the total 36 cows, 2 remain. One, as you know, belongs to me anyway; the other, well, I think it fair that I also keep, as I have divided your inheritance in a manner that is fair for all.*

Papa was speechless because that's exactly the answer to the mathematical riddle. He was pleased and very proud of my accomplishments. However, I have a great deal more to learn. If it took centuries for mathematics to develop, I can just imagine how many more concepts there are that I do not yet know.

Tuesday—December 15, 1789

The foundation of algebra was established in ancient Egypt and Babylon hundreds of years ago. Of course, the ancient mathematicians did not think in terms of equation as we know it today. For example, Euclid developed a geometrical approach that involved finding a length, which represents the root $x$ of a quadratic equation. He had no notion of equation or coefficients; Euclid worked with purely geometrical quantities. Also, the Babylonians developed an algorithmic approach to solving problems, which could be interpreted as the quadratic equation we know now. Their solutions were always positive quantities, since they represented lengths and areas, just like Euclid's approach. That is why negative numbers were not known in antiquity.

The word "algebra" is derived from the Arabic word for "restoration," *al-jabru*. Historians find that Arab mathematicians knew the solutions of equations as the "science of restoration and balancing." Ancient mathematicians wrote out algebraic expressions using only occasional abbreviations, but by medieval times Islamic mathematicians were talking about arbitrarily high powers of the unknown $x$, and they worked out the basic algebra of polynomials, without using modern symbolism, of course. A Latin translation of Al-Khwarizmi's *Algebra* appeared in the 1100s. A century later, the Italian mathematician Leonardo Fibonacci achieved a close approximation to the solution of the cubic equation $x^3 + 2x^2 + cx = d$. Fibonacci traveled extensively, so he probably learned and used an Arabic method of successive approximations in one of his trips.

Early in the sixteenth century, the Italian mathematicians Scipione del Ferro, Niccolò Tartaglia, and Girolamo Cardano solved the general cubic equation in terms of the constants appearing in the equation. Ludovico Ferrari, who was a pupil of Cardano, soon found an exact solution to equations of the fourth degree, and as a result, mathematicians for the next several centuries tried to find a formula for the roots of equations of degree five, or higher.

Also in the sixteenth century, algebra was further developed with the introduction of symbols for the unknown and for algebraic powers. The French philosopher and mathematician René Descartes introduced such symbols in his Book III of *La Géométrie*. Descartes' geometry treatise also contains theories of equations, including the rule of signs for counting the number of what Descartes called the "true" (positive) and "false" (negative) roots of an equation.

Many mathematicians developed algebra in a period of many centuries. Each one made a contribution to the current knowledge, like adding a gold coin to a treasure chest. Right now it looks plentiful, glistening in the sun. However, the chest of knowledge is not full. Mathematics, I am sure, still holds many mysteries that need to be discovered. That is why I wish to become one of the mathematicians who will contribute to the treasure chest of knowledge.

Friday—December 25, 1789

What a wonderful day! Starting after morning prayers, everybody was in a joyful mood, hugging and kissing, wishing one another good things. My sisters and I helped Maman set up *La Crèche*, the Christmas crib, with figurines representing the shepherds, the farm animals, and the holy family. The three wise men will be added to the nativity at Epiphany in January.

Early in the day, we had a special Christmas lunch. We started with fresh vegetables and *anchoiade*, a salty dressing made with anchovies that I dislike. We ate a spinach omelette, fish with tomato sauce, and goat cheese. We had the thirteen traditional Christmas desserts, fruits prepared in different ways, and Provençal wine. Angelique asked, as she does every year, why there are thirteen. So Maman explained again that the thirteen desserts represent Jesus and the twelve apostles.

And as she does every year, Maman placed three candles on the table to represent the Holy Trinity. I know this is ridiculous, but Madame Morrell tied the corner of the tablecloth with a knot so that, according to her, the devil can't climb on the table. After lunch Mme Morrell insisted on keeping the table set until after the Christmas Mass, so that "the spirit of the Saints can eat what is left." Although we do not believe in those silly superstitions, my parents do not object; like other harmless customs, they are part of the traditional festivities of people from the villages.

After lunch we visited friends and relatives. For dinner, Mme Morrell prepared mutton, lamb chops, asparagus soup, boiled vegetables, and cheese. Then we dashed to *La Pastorale*, the shepherd's play, in the courtyard of Notre Dame. The performance finished in the church right before midnight Mass. The ceremony was beautiful; there were hundreds of candles with flickering flames casting a golden glow on the statues of saints and virgins. The smell of incense was overwhelming. The music from the church organ made me shiver and more so when the chorus sang.

The ritual ended past midnight. Upon leaving the church, we drank a sip of warm wine Papa bought from an old woman by the bridge. The night was cold when we returned home, but it did not bother me. During the ride home, a pale moon cast its shimmering light on the city. I looked up and discovered a dark sky full of blinking stars. How happy it made me feel. At home we gathered at the dinner table once more to express good wishes to each other.

C♋ЄꝊ

Hôtel de Ville, city hall of Paris, center of politics and government.

The siege of the Bastille, July 14, 1789.

Arrival of the royal family in Paris after the assault in Versailles, October 6, 1789.

*Paris, France*
*Year 1790*

# 2
## Discovery

I SIT IN PAPA'S STUDY, watching through the window the velvety white snow blanketing the city. The afternoon light intermingles with the shadow of the night. The street lamps are lit, adding a golden glow to the twilight. People go about their affairs, poor and rich, conscious of each other's social status. The wealthy travel in their fancy carriages, bundled up in fine furs, while the less fortunate trudge in the muddy streets, wearing worn-out garments to protect them from the cold wind. Rich and poor, young and old, humans are alike in their desires, but they are not socially equal.

I never thought about it before, but now it's plain to see that the citizens of France do not share equal rights under the law. Last year the workers and poor peasants revolted, unhappy with the unfair tax structure, and the financial crisis that led to increase in prices and scarcity of goods, especially bread. While the royal family and the court were living in luxury at their Palace in Versailles, the peasants and the poor workers were suffering.

Papa and his friends criticize Louis XVI for being an incompetent ruler. The king is good, has a noble heart, but lacks leadership and is not in touch with his subjects, especially the lower classes. Papa asks how is possible that a feeble monarch who seeks only to satisfy his own pleasures can govern a nation? And many people dislike Marie Antoinette because she was capricious and irresponsible in her early years as queen. Rumors about her abound, that she spends huge amounts on clothes while the poor people starve outside her palace. Other people loathe the queen simply because she is Austrian.

Now I also understand why the people forced King Louis and his court to move to Paris. They hope the king will be more effective

61

helping his subjects. And he is trying. Louis XVI is attempting to solve the financial crisis by removing some of the exemptions on taxes. The aristocrats are furious. His Majesty faces strong opposition from the nobility. That is the reason he called the meeting of the Estates General last year. However, very little has changed for the poor people since then. Even after moving to the Tuileries Palace in Paris, the king cannot resolve the social conflicts and the financial crisis of the nation.

What will happen next? Will 1790 be a better year? I close my eyes and make a wish, as I used to do when I was a little girl. May the New Year bring us peace.


Wednesday—January 6, 1790

Winter is my favorite season, perhaps because the city becomes quiet, as if life stands still. The white snow covering the roofs makes the buildings look enchanted. In winter I spend more time in solitude and study. In the coldest days, I huddle in my father's library in front of the fireplace. This evening, to entertain my sister I made up the following story:

*There was once a king who owned many of the most gorgeous palaces ever built. He also had several daughters who one day would rule the immense kingdom. When he died, the monarch left the palaces to be divided among all his daughters. In his testament, the father stated that the division must be as follows: the oldest daughter would receive one palace and a seventh of the remaining buildings. The second daughter would get two palaces and a seventh of what remained. The third girl would receive three palaces plus a seventh of the rest. And so on until all his daughters received their share of the land.*

*The princesses were unhappy, believing some were getting less than the other sisters. They went to their mother crying. The queen, who was an intelligent woman, responded that their protest was unfounded, as their father had given all his daughters a fair share of the kingdom.*

*Can you tell how many princesses the king had, and how many palaces each one inherited?*

Angelique guessed the answer, but incorrectly. This is how one can solve the riddle:

The first princess received 1 palace and a seventh of 35, which is 5. That is, she received 6 palaces, leaving 30 for the rest of his sisters.

Of the 30 palaces, the second princess got 2 plus a seventh of 28, which is 4. She was given 6 palaces, leaving 24 for his younger sisters.

One can go on doing this analysis in the same manner until all palaces are given. So the correct answer for the riddle is: the king had 6 daughters and left them 36 palaces to be divided equally.

The problem can be solved with algebra, stating it with a simple formula that has two unknowns. Let the number of palaces be $x$, and the number of princesses $(n - 1)$, and one can derive the following formula:

$$x = (n - 1)^2$$

I determine that the first princess would receive 1 palace and $1/n$ of the rest. The second would get 2 palaces plus $1/n$ of the rest. And so forth. One can prove it by solving for the case in which $n = 7$, so that $x = 36$.

Monday—January 11, 1790

$I$t's so cold right now that the ink is freezing in my pen. I must put away writing and get under the warm bedcovers. I will read a new book before going to sleep.

$\text{T}$here is a special equation nobody knows how to solve. At first glance the equation appears very simple: $x^n + y^n = z^n$. However, not even the best mathematicians in the world know how to solve it. This is a Diophantine equation, named after Diophantus of Alexandria. There is no general method for solving such equations. What makes the equation more fascinating is a mysterious note that a mathematician wrote in the margin of a book before he died. His name was Pierre de Fermat, a Frenchman who lived about one hundred and thirty-five years ago.

Monsieur de Fermat claimed that the equation has no non-zero integer solutions for $x$, $y$, and $z$, when $n > 2$. After his death, his son found his father's *Arithmetica*, the book written by Diophantus hundreds of years ago—a book I am studying. The son discovered that in the margin of one page Fermat had written: "To divide a cube into two other cubes, a fourth power or in general any power whatever, into two powers of the same denomination above the second is impossible, and I have assuredly found an admirable proof of this, but the margin is too narrow to contain it."

Fermat meant that there are no whole number solutions for equations like these: $x^3 + y^3 = z^3$, $x^4 + y^4 = z^4$, $x^5 + y^5 = z^5$, and so on. These are equations of the general form $x^n + y^n = z^n$. I've solved the equation $x^2 + y^2 = z^2$ many times. Using the Pythagorean Theorem to solve the triangle with two sides $x$ and $y$ equal to 3 and 4, respectively, yields the sum of perfect squares: $3^2 + 4^2 = 5^2$. Well, this equation is easy.

But Fermat claimed that when the exponent $n$ is greater than 2, the equation $x^n + y^n = z^n$ has no solutions! It is very difficult to prove this statement because there are an infinite number of equations and an infinite number of possible values for $x$, $y$, and $z$. A great Swiss mathematician named Leonhard Euler obtained only a partial proof for the case $n = 3$. A full proof would require the inclusion of all cases to $n$ infinite.

Would it not be wonderful that I could prove one day Monsieur Fermat's claim?

Mother is so happy. She met Queen Marie Antoinette at the Tuileries Palace. Not by herself, of course. Maman, Mme Fougeret, and other ladies in the committee of the *Charité Maternelle* had an audience with the queen. They asked for financial support to help poor unwed mothers and orphan children.

Maman is glowing with pride. She says that Her Majesty is a very compassionate and kind lady and agreed to support the charity. Maman thinks that the only reason some people blame the queen for the financial problems in France is because Marie Antoinette looks like she is spending all the wealth on fancy stuff. Maman likes her very much and doesn't believe all the bad rumors against her.

As the daughter of the Empress Maria Therese of Austria, Her Majesty was brought up believing her destiny was to become queen. And she did. At fourteen she married the crown prince of France. That was before I was born, in 1770. Four years later, she became queen of France when her husband was crowned King Louis XVI.

Maman thinks that the stories of the queen's excesses are vastly overstated. Even Papa acknowledges that, rather than ignoring France's growing financial crisis, Marie Antoinette reduced the royal household staff, eliminating many unnecessary positions that were based on privilege. Precisely because of this, the queen offended the nobles, adding their condemnation to the scandalous and false rumors spread by disloyal subjects. Maman insists that the aristocrats are the ones who reject the financial reforms the government ministers have proposed; the king favors social changes.

Now many political leaders are inciting the citizens against the monarchy. They will not appreciate the good deeds of Her Majesty. Maman, however, will always remember Queen Marie Antoinette with affection. She spent the evening telling us how beautiful and gentle she is, how regal her demeanor, and how graciously she received the ladies in the Tuileries.

The best time for me to study is at night, after everybody goes to sleep. But twice a week I study in Papa's library when my parents leave to attend the salons of Mme de Maillard and Mme Geoffroy. The salons are intimate gatherings at which people meet and talk about philosophy, politics, literature, and many other topics of current interest. During those evenings my parents have supper there, and they do not return home until late at night. Thus, I can study in the library where is warm and cozy; Mme Morrell keeps the fireplace burning until Papa goes to bed.

Tonight is freezing cold, and instead of going out, my parents decided to play cards in the library, thwarting my plans to study there. Just now Millie came to bring me a cup of hot chocolate and whispered that everyone went to bed. It means I can steal a little time for my calculations without interruptions.

What would I do without Millie? Especially now that's bitterly cold. Mother ordered Madame Morrell to stop stocking my heater after eleven o'clock, and to ration my candles. If Mother found out that Millie sneaks out in the middle of the night to bring me a candle or a cup of chocolate, the two of us would be in trouble. It is one thing to help me with the embroidery, but another to go against my mother's orders. I must search for candles myself because it would not be fair to have Millie punished for helping me.

I am so glad I have my lap desk because I can write my notes bundled up under the blankets. The hot cup of chocolate helps keep my hands warm to hold the pen to write.

Sunday—February 7, 1790

I found a rare book I can't read. It is about mathematics, but it is written in Latin, the universal language of scholars. Papa explained that scholars write in Latin so that their work is understood by others anywhere in the world. He added, "And you, *ma petite élève*, must know Latin if you wish to learn from this book." Papa knows I like challenges, and if he believes I can do it, then I will teach myself the language of mathematicians.

Even though I could not understand the words, I kept perusing the book until I found an equation that caught my attention. The equation looks simple, yet elegant and pure: $x^3 + 1 = 0$. So I copied it in my notebook assuming it'd be easy to solve. Then I discovered that there are three values of $x$ that make the left side of the equation 0. For sure, $x = -1$ is one solution since $(-1)^3 + 1 = -1 + 1 = 0$, but what are the other two solutions? It can't be $x = 1$ because I would get: $(1)^3 + 1 = 2$. And it can't be any number greater than 1 either, so what could the other two values of $x$ be? I wish I could read Latin because maybe the solution is explained with words.

Oh, my candle is almost exhausted. It is freezing, and the ambers in the heater are already cold. My hands feel like ice and I cannot hold the pen any longer. I better get some sleep and tomorrow I'll figure out how to solve this problem.

Sunday—February 14, 1790

Sir Francis Bacon said: "*Ipsa scientia potestas est,*" which means, "Knowledge itself is power." Bacon was an English philosopher who argued that the only knowledge of importance to man is empirically rooted in the natural world. Bacon believed that a clear system of scientific inquiry would assure man to master the world.

Indeed, knowledge is power. I need to learn the language used by scholars to write about their discoveries. Of course, it is not easy to learn Latin by myself. I have to look up almost every word in the dictionary just to understand a bit of what I am reading. I guess the meaning of some words, but I have a hard time understanding an entire sentence. Translating takes so much time!

It occurred to me that I could take lessons from the nuns at the convent, since they are fluent in Latin. So I asked Maman, without revealing that it would help my study of mathematics. I expected that my mother would be most favourably disposed since this knowledge could be considered part of my religious education. Indeed, Maman promised to speak with the Mother Abbess at the Sisters of St. Joseph convent. I was delighted and couldn't wait to start my tutoring. But at dinner, Papa told me to wait until things settle down. He remarked that the Assembly just voted to suppress religious orders and monastic vows. Beginning today the government will give the nuns a choice of either leaving the convent and accepting government pensions, or remaining in especially designated monasteries chosen by the State. It's still too early to know what the Sisters at St. Joseph would do, so for now my lessons in Latin will have to wait.

Très bien. I'll continue learning Latin on my own until we find out. If it gets difficult, I'll tell myself Nil desperandum, which means "do not despair," and I will keep trying.

Saturday—February 20, 1790

I am studying equations that can only be solved with *imaginary numbers*. Imaginary numbers are the product of a real number and the square root of −1. An imaginary number is a multiple of a quantity called "*i*," which is defined by this property: $i$ squared equals −1, or $i = \sqrt{-1}$. These numbers are called imaginary because they did not fit the definition of number that was available when first introduced. Imaginary numbers arose from the need to solve certain quadratic equations that many mathematicians believed had no solutions.

At first, I could not think of a number having a negative square root. Then, I came up with a very simple way to see imaginary numbers and conclude that many problems in mathematics could not be solved without them. For example, if I have to solve the quadratic equation $x^2 + 1 = 0$, that means that $x^2 = -1$, thus the solution is $x = \sqrt{-1}$. Obviously, without imaginary numbers I would not know what $\sqrt{-1}$ is, and thus I could not arrive at the solution.

With the imaginary quantity $i$, and letting $a$ and $b$ be positive or negative real numbers, I can construct infinitely many numbers of the form $a + ib$ so that I can find the values of $x$ that make my equation true; the two solutions to the equation $x^2 + 1 = 0$ are $i$ and $-i$, since $x^2 = -1$, or $x = \pm\sqrt{-1} = \pm i$. These solutions are complex imaginary. As I said, the imaginary unit is denoted and commonly referred to as $i$. Although there are two possible square roots of any number, the square roots of a negative number cannot be distinguished until one of the two is defined as the imaginary unit, at which point $+i$ and $-i$ are obvious. Since either choice is possible, there is no ambiguity in defining $i$ as "the square root of $-1$."

Ancient mathematicians thought that it was impossible to take the square root of a negative number. That's because they had not thought of numbers that were negative when squared. The first mathematicians had only real numbers at their disposal, and thus didn't know how to take the square root of a negative number.

Imaginary numbers were invented a few hundred years ago. I say "invented" because all numbers are a creation of our minds. People create numbers to help them solve equations; numbers are not physical objects! So, mathematicians invented imaginary numbers because they needed them, just like they invented many other mathematical concepts.

I wonder what new mathematics will be revealed to me now that I know this number called $i$.

Friday—February 26, 1790

I am terribly upset with my sister Angelique. This afternoon, when I was in the library, she came into my bedroom and rummaged among my things. She opened the drawer in my lap desk and found my papers. Silly Angelique—when she saw my writing about imaginary numbers she ran to tell Mother that I was going crazy. She started teasing me, moving an imaginary pen in the air, saying, "O la la, Sophie plays with imaginary numbers!" Mother dashed in and demanded that I throw away my notes, saying that this was the reason she didn't approve my study of mathematics. I tried to explain that imaginary numbers are not a figment of my imagination, but Maman wouldn't listen. She took all my notes and was about to burn them.

At that precise moment, Papa appeared and asked what the ruckus was about. After I told him, he smiled sympathetically and assured my mother I am not losing my mind. He persuaded her to return my notes. Maman did it reluctantly. First, she made me promise that I won't spend all my time in "this ridiculous pursuit unbecoming of a young lady." That is what my mother said, among other things I do not want to repeat. Why can't she understand that learning mathematics is not a silly pursuit? It is the only thing that matters to me!

My mother and my sister made such a fuss, thinking that my study of imaginary numbers is abnormal. Of course I am not crazy, but it would be futile to try to explain it to them. However, I should be patient and not be so hard on them for not understanding. It took centuries for these numbers to become fully understood and used.

It all began with the discovery of negative numbers. Before the seventeenth century, many mathematicians didn't know or accept negative numbers. Even great mathematicians like Blaise Pascal resisted this idea. Pascal was a French philosopher and physicist, a colleague of Pierre de Fermat, who lived during the reign of Louis XIV. I read that Pascal once said: "Who doesn't know if you take 4 away from nothing, you're still left with nothing?" Pascal knew about negative numbers because, a century before, the Italian mathematicians Niccolo Fontana Tartaglia and Girolamo Cardano discovered negative roots when trying to solve cubic equations. These negative roots eventually led to the invention of imaginary numbers.

Cardano was solving cubic equations such as $x^3 = 15x + 4$, when he obtained an expression involving the square root of $-121$. Cardano thought that he could not take the square root of a negative number, however, he knew that $x = 4$ is one solution to the equation. But he was unsure about the other two solutions. He wrote to Tartaglia asking for his opinion on this puzzling answer, but Tartaglia was unable to help him. Cardano almost discovered imaginary numbers when he concluded that the problem of dividing

10 into two parts, so that their product is 40, would have to be $5 + \sqrt{(-15)}$ and $5 - \sqrt{(-15)}$.

Cardano wrote a book entitled *Ars Magna* (*The Great Art*) where he included negative solutions to equations, but he called them "fictitious" numbers. In his book, Cardano also noted an important fact connecting solutions of a cubic equation to its coefficients, namely, that the sum of the solutions is the negation of *b*, the coefficient of the $x^2$ term. A few years later, another Italian mathematician named Raphael Bombelli gave several examples involving these new numbers, and he gave the rules for their addition, subtraction, and multiplication.

René Descartes introduced the terms "imaginary" and "real" in his book *La Géométrie*. Descartes wrote: "Neither the true roots nor the false are always real; sometimes they are, however, imaginary; namely, whereas we can always imagine as many roots for each equation as I have predicted, there is still not always a quantity which corresponds to each root so imagined. Thus, while we may think of the equation $x^3 - 6x^2 + 13x - 10 = 0$ as having three roots, yet there is just one real root, which is 2, and the other two, however, increased, diminished, or multiplied them as we just assigned, remain always imaginary." Finally, in 1777 Leonhard Euler recommended the general use of these imaginary numbers in his book *Introduction to Algebra*.

I just solved Descartes' cubic equation $x^3 - 6x^2 + 13x - 10 = 0$. In addition to the solution $x = 2$, the other two roots are $x = (2 + i)$ and $x = (2 - i)$. Without imaginary numbers, I could not have found the square root of the negative numbers. How wonderful! It seems as if a new door to mathematics is open to me now that I know imaginary numbers.


Tuesday—March 2, 1790

My father asked me how to compound interest. He lends money and uses the simple formula $I = P \cdot R$ to compute the interest for each loan ($I$ = interest, $P$ = principal or original amount loaned,

71

and $R$ = rate of interest). Now Papa wants to compound interest to charge not only for the original loan, but also charge for some of the interest owned. He wants to charge interest on the interest because some people take too long to pay him back.

I found the formula $A = P\,(1 + R/n)^{nt}$ to calculate the compound interest, where $A$ is the total amount he would be paid, $t$ is the elapsed time, and $n$ is the number of times he can compound the interest in the time $t$. Papa wants to figure out how many times he can compound the interest in order to get the highest return for his loans.

Since Papa did not tell me the interest he charges for each loan, I assume it is 100% per year, and further assume he loans one livre (1 $l$). These values might not be realistic, but this assumption will simplify the analysis.

Thus, with $P = 1\ l$ and $R = 1.00$, and $t = 1$, I get the following:

Compounding annually (once a year), $n = 1$ and $A = 1\ l\,(1 + 1.00/1)^{1\times1} = 2.0\ l$

Compounding biannually (twice a year), $n = 2$ and $A = 1\ l\,(1 + 1.00/2)^{1\times2} = 2.25\ l$

Compounding quarterly (four times a year), $n = 4$ and $A = 1\ l\,(1 + 1.00/4)^{1\times4} = 2.44\ l$

Compounding every month, $n = 12$ and $A = 1\ l\,(1 + 1.00/12)^{1\times12} = 2.61\ l$

I keep increasing $n$, doubling and tripling it, but $A$ grows at a very slow pace. Even if the interest is compounded daily, I do not see much growth. In fact, the total amount of the payment is just $A = 1\ l\,(1 + 1.00/365)^{1\times365} = 2.715$!

There appears to be a limit to the earnings from the compound interest formula. I have to tell Papa that he could compound the interest every day, but the most earning he would net is limited to about 2.715. So, even if the interest were compound every hour, he still would get less than three times the amount he loans. Amazing! I wonder why?

72

Mathematics and the game of chess are connected. Today I learned how. I was in the library, playing chess by myself, when Papa arrived with Monsieur de Maillard. He startled me when he moved a piece on the board and commented on my strategy. As they settled in armchairs to drink coffee, M. de Maillard told me an extraordinary tale of the origin of this game. I'll attempt to recount the story.

Many centuries ago, there was a king who wanted a unique game that no one else had, a game that was so original that it could be played over and over in endless combinations, a game that would teach his children to become better thinkers and better leaders on the battlefield. With this in mind, a wise man invented the game that we now call chess. The monarch was very pleased, and he offered as a payment anything the man would want—gold, jewels, anything. The wise man asked to be paid with wheat grain. When asked how much, the man replied that he wanted the amount based on the number of squares on the game board.

His formula was simple; for square one, he wanted one grain. For square two, it would be doubled to two pieces, for square three, it would be four pieces, for square four it would be eight pieces, and so forth until all sixty-four squares were filled in this proportion. The king was surprised by the seemingly modest request, but ordered a sack of wheat grain to be brought. The servants patiently began to place the pieces of wheat on each square on the board as requested by the wise man. But the grain ran out and more sacks were brought up. Soon, they discovered that not even all the grain in the kingdom would be sufficient to cover half of the squares on the chessboard.

M. de Maillard asked me if I could determine how much grain would be needed. I didn't know offhand, but I scurried to my room and decided to give it a try. I started with numbers to illustrate the distribution of wheat grain on the chessboard using numbers. Since there are eight squares on the first row, I write:

1    2    4    8    16    32    64    128

in the second row:

256   512   1024   2048   4096   8192   16,384   32,768

in the third row:
65,536   131,072   262,144   524,288   1,048,576   2,097,152   .....

| 1 | 2 | 4 | 8 | 16 | 32 | 64 | 128 |
|---|---|---|---|---|---|---|---|
| 256 | 512 | 1024 | 2048 | 4096 | 8,192 | 16,384 | 32,768 |
| 65,536 | 131,072 | 262,144 | 524,288 | | | | |
| | | | | | | | |
| | | | | | | | |
| | | | | | | | |
| | | | | | | | |
| | | | | | | | |

*Très bien*! Enough!

I have to stop here. Now I just imagine the huge numbers on the squares of the fourth row and the remaining squares on the chessboard. Without doing the calculation I know that on each square I can represent the number of grain by the number $2^n$, with the first square $2^0 = 1$, and the number on the last square $2^n = 2^{63}$. Then I have to add the grain on all squares, making the total number of grain enormous. Even $2^{63}$ is such a huge amount of grain that I cannot visualize it!

How can I tell this story using only mathematics? I could write a sum $S = 1 + 2 + 4 + 8 + 16 + 32 + \ldots + 2^{63}$, which gives me an idea of the total number of grain. That's all I can think right now.

Monday—March 15, 1790

$T$oday was cold and gray, and the rain made me feel gloomy. After dinner, we gathered in Papa's library around the cozy fireplace. Angelique was restless and cranky so I offered to teach her how to play chess. At first she was not that interested, but then I told her about the monarch who wanted a game to teach his children to be smarter, and she sat listening quietly. My sister likes fairy tales, so it was easy to embellish the story by creating more characters that she found interesting, although none of them had anything to do with chess.

In my tale the wise man became a handsome prince; he would marry the king's beautiful daughter after winning the game. I also added that the ruler was very pleased with the new game, and thus offered, as a dowry for the lovely princess, anything the smart prince would want—gold or diamonds. I also changed the part of the story where the wise man asked to be paid with wheat grain. To keep Angelique's interest, I changed the story slightly by saying that the prince asked for diamonds! My sister was enchanted and was more interested in learning to play, imagining perhaps the sparkling diamond on each square of the chessboard.

We spent a couple of hours playing chess, and I explained the method to determine the number of grains (or diamonds) that would result from doubling the number in each subsequent square of the chessboard. I don't think Angelique grasped what I was telling her, but she listened patiently. The number of grains on the chessboard would be a sum of 2s raised to increasing powers, from $n = 0$ to $n = 63$: $S = 2^0 + 2^1 + 2^2 + 2^3 + 2^4 + 2^5 + \ldots + 2^{63}$.

More compactly, $S = 2^{64} - 1$ grains. Hence, with about $100 = 10^2$ grains in a cubic meter, the total volume of grain would be nearly two hundred thousand million cubic meters! That would be a lot of wheat indeed!

| $1$ | $2^1$ | $2^2$ | $2^3$ | $2^4$ | $2^5$ | $2^6$ | $2^7$ |
|---|---|---|---|---|---|---|---|
| $2^8$ | $2^9$ | $2^{10}$ | $2^{11}$ | $2^{12}$ | $2^{13}$ | $2^{14}$ | $2^{15}$ |
| $2^{16}$ | $2^{17}$ | $2^{18}$ | $2^{19}$ | $2^{20}$ | $2^{21}$ | $2^{22}$ | $2^{23}$ |
|  |  |  |  |  |  |  |  |
|  |  |  |  |  |  |  |  |
|  |  |  |  |  |  |  |  |
|  |  |  |  |  |  |  |  |
|  |  |  |  |  | $2^{61}$ | $2^{62}$ | $2^{63}$ |

Thursday—March 25, 1790

$M$other does not appreciate me! She doesn't understand why I study mathematics. She can't! This evening I overheard her telling Papa that I am taciturn and self-absorbed. Mother complained that there are times when she asks me a question, and I do not even lift my head up from the book I am reading. She confided that, "at times Sophie sits there with her eyes glazed over, looking away like in a trance." I do not know what Papa replied after that. I did not want to hear anymore so I left as quietly as I had arrived.

I am hurt and upset. Why does my mother call me self-absorbed? Does she not understand that mathematics requires deep thinking and full concentration? When I am engaged in the solution of a problem, or the study of a new topic, I submerge myself in it, leaving behind everything else. This should not worry her.

My dear mother; she was raised following the Rousseauistic image of proper womanly conduct. She believes, like many women, in the idea of feminine fulfillment advocated by the French philosopher Jean-Jacques Rousseau, that women should devote themselves to the happiness of their husbands and children. Maman thinks that women should remain silent and not express their opinions, especially if they are in conflict with their husband's point of view. She believes that a girl should learn domestic skills to become a good wife and mother, not a scholar. I disagree but can't contradict her. I just wish she'd accept me for what I am.


Holy Thursday—April 1, 1790


I witnessed a memorable sacrament at church. For the first time in history the *Pedilavium*—washing of the feet— was performed by King Louis and Queen Marie Antoinette. The ceremony began when twelve of the poorest people, dressed in new clothes donated by the monarchs, walked to the front of the church. They sat on a bench with their right foot bare and rested it on the edge of a basin of water. His Majesty approached them and "washed" each person's foot by pouring water over it with a scoop. The queen followed, placing a white napkin over the wet foot. It was all very solemn; people watched with admiration and awe because this is the first time a king of France has performed such a humble act. It's the king's way to gain his subjects back and ease the social tensions of the past year.

I turned fourteen today but nobody said a word about it. We spent the day in solitude and prayer. On this day people remember the trial of Jesus Christ that led to his crucifixion. During this holy week we fast and pray in preparation for Easter. Fasting is not difficult for me because it just means that we eat very little every day, mainly bread and water, but my little sister Angelique still struggles with it, for she can't abstain from eating desserts. By evening, Madame Morrell fixes her some bouillon because she is starving. Millie also gave her cheese and a bit of jam. Then Angelique asks me whether God will be mad at her for not fasting.

I hear Mme Morrell extinguishing the lamps in the hall and Millie climbing the stairs to their quarters. The midnight bells will toll very soon so it's time to snuff my candle.

Easter Sunday—April 4, 1790

Today was cold, but the sun was out, shinning brightly over the clear blue sky. After Easter Mass we rode to the Tuileries gardens. The pathways in the gardens were crowded; women were parading with their newest dresses and feathered hats, celebrating the new beginning that Easter represents.

As we strolled by the Tuileries main entrance, we recognized the new Dauphin, the heir to the throne, playing with his sister and their governess. The boy must be five years old now, very friendly and precocious. The princess, a blond girl who is about my little sister's age, was aloof and hardly looked at the people. Angelique waved at the royal children, and the Dauphin waved back enthusiastically. My sister was very proud; although the boy was waving at all the people watching him play in the palace's garden. Maman says that the Tuileries is not as sumptuous as Versailles, so the royal family had to adapt to a most modest home. The little boy, however, did not seem to mind. He looked as happy as any other child his age.

Papa is going to purchase some land located in the north side of Paris. He will buy it at the auction of properties that belonged once to the Church of France. Now the government is selling it in order to pay for the national debt. Papa asked me to estimate how much land he can buy with 2000 livres. This is quite a trivial problem, mathematically. But for somebody trying to buy property it is very important.

For example, if the lot were perfectly square, its area would be the product of the length of one side by itself, that is, $x \cdot x = x^2$. With $x^2$ representing an area, I write the relationship between the area and the cost. For example: $x^2 = 2000$. Solving this equation, I get the length of the lot, $x$ = square root of 2000 ≈ 44.7 m. With 2000 livres Papa can buy a square lot that measures 44.7 × 44.7 meters. Now, if Papa also has to buy a fence to enclose the land, the formula must include both area and length. The length of fence is simply the perimeter of the square lot, or $4 \cdot x$. Now I assume that Papa has 2021 livres to purchase everything. In this case, I write a relation $x^2 + 4x = 2021$. To solve this quadratic equation I note that $(x^2 + 4x)$ represents the first two factors of the binomial $(x + 2)^2$, which expanded is $(x + 2)^2 = x^2 + 4x + 4$. Thus, I can use this fact to write my equation as

$$x^2 + 4x + 4 = 2021 + 4$$

(I just added a 4 to both sides of the equation.)

Simplifying, the previous equation becomes

$$x^2 + 4x - 2021 = 0$$

This is a quadratic equation of the form $ax^2 + bx + c = 0$. Real solutions exist only when $b^2 - 4ac \geq 0$ and are given by

$$x_1 = \frac{-b + \sqrt{b^2 - 4ac}}{2a} \text{ and } x_1 = \frac{-b - \sqrt{b^2 - 4ac}}{2a}$$

if $b^2 - 4ac = 0$, the two solutions are identical, that is $x_1 = x_2$.

Now, since this is a problem of lengths and areas, the root of the quadratic equation has to be real. In other words, in this case $x_2$ would be meaningless. Thus, I only use the first root:

$$x = \frac{-b + \sqrt{b^2 - 4ac}}{2a} = \frac{-4 + \sqrt{16 - 4(1)(-2021)}}{2} = \frac{-4 + \sqrt{8100}}{2} = 43$$

So, with 2021 livres, Papa could buy a 43×43 meter square lot and the fence around it.

*Très bien.* Now, what if the land is not perfectly square? I could assume, for example, that the lot is rectangular. In such a case, the area is equal to $x \cdot y$, and the perimeter is equal to $(2x + 2y)$. My equation would then be: $x \cdot y + 2(x + y) = 2021$. How can I solve it? Easy! All I need is a relationship between $x$ and $y$. Let's say that $y = 3x$, then I can substitute as follows:

$$x \cdot y + 2(x + y) = x(3x) + 2[x + (3x)] = 3x^2 + 2x + 6x = 2021$$

$$3x^2 + 8x - 2021 = 0$$

Once again, I have a quadratic equation that I can solve as before. So, no matter what geometric shape I am given, as long as I can find a relationship between the sides $x$ and $y$, I will arrive at a solution.

Friday—April 9, 1790

An unexpected surprise for me today! "Letter for Mademoiselle Marie-Sophie Germain," an old gentleman at the door announced gravely. We were sitting in Mother's study, ready for our writing lessons, when Millie opened the door followed by a stranger. He introduced himself as the valet of Monsieur the Marquis de Condorcet. I was unsure. Did he call my name?

All eyes turned to me at once, stumped as I was. After clearing his throat, the gentleman repeated, "I bring this communication for Mlle Sophie Germain on behalf of the Marquis de Condorcet." With a nod, Maman encouraged me to stand up and accept the letter, which the old man handed to me along with a package. I was stunned, but I managed to curtsy and thank him. My sisters rushed to me after the man left, and Maman urged me to read the missive and to open the package. They were excited, but not as much as I was.

There was no mistake; the letter was addressed to me! My name was written elegantly in black ink letters on the ivory parchment. I broke the aristocratic wax seal with trembling hands.

The message was short: *May you find Euler's memoir stimulating. It is for the few who truly appreciate the beauty of mathematics.* Signed "Jean-Antoine-Nicolas Caritat, Marquis de Condorcet." My hands trembled in anticipation, and Angelique had to contain herself from removing the wrapping for me. The package contained two volumes with leather covers of the book *Introductio in analysin infinitorum* by Leonhard Euler, published in 1748. Angelique quickly lost interest when she saw mathematics and the text written in Latin.

Maman was unsure that I should accept a gift from such a distinguished gentleman. She is pleased with the honor, of course, but she does not understand why a man would encourage my study of mathematics. She cannot prevent me from reading these books, but after dinner she reminded me to go to bed early. Which I do, and she knows it. Maman also made me promise to write a nice letter to the Marquis to thank him for his generosity.

I am thrilled to receive this unexpected but extraordinary gift. Now I understand why I am so lucky. Papa introduced me to the Marquis recently, when we're at the bookshop near the Louvre. While exchanging pleasantries, Papa mentioned my interest in mathematics. M. Condorcet beamed, adding that he, too, is interested in the most beautiful of sciences. After we said goodbye, Papa praised M. Condorcet, stating he is a man of considerable erudition, a great mathematician at the French Academy of Sciences.

M. Condorcet must think I need to study this book. Even if I do not understand every sentence I read, I can study the equations. So far, I translated the title of the book as "Introduction to Analysis of the Infinite." M. Condorcet's excellent gift means more to me than all the gold in the world. I will tell him so in my letter.

My books on the analysis of the infinite wait on my table, beckoning me to begin the journey of discovery and learning.

Friday—April 16, 1790

Euler's *Introduction to Analysis of the Infinite* opens with the definition of *functio,* a Latin word that means "function." Euler defined a function as a variable quantity that is dependent upon another quantity. For example, the analytic expression $y = 2x^2$ is a function, written as $f(x) = 2x^2$. Thus, a function is simply an association between two or more variables; to every value of each independent variable corresponds exactly one value of the dependent variable, in a specified domain where the function is valid. When I write the expression $f(x) = 2x^2 + 7$, for each value of $x$ I get a value of the function $f(x)$; so $f(x)$ denotes the dependent variable, and $x$ is the independent variable.

An equation in which the dependent and independent variables occur on one side of the equality sign is said to be in implicit form. For example, the expression $2x^2 - 2y = 6$ is written in implicit form, but it can be written in the more clear explicit form, $y = x^2 - 3$, or also as $f(x) = x^2 - 3$.

The domain of the function is the collection of all values that the independent variable can take. For example, given the function $f(x) = 1 - x^2$, and defining the independent variable $x$ as a real number, the domain of this function must be the collection of all real numbers from $-\infty$ to $\infty$, where the symbol $\infty$ represents "infinite." This is because the numerical value of $x^2$ will always be zero or positive. As $x$ goes toward infinite, $\infty$ or $-\infty$, the function $f(x)$ approaches minus infinite, or $f(x) \to -\infty$. Of course, $f(x)$ will reach its greatest value when $x = 0$. The range of this function includes the number 1, and all the real numbers less than 1. Written with mathematical notation, the range is given by $-\infty < f(x) \le 1$.

There are many types of functions. Linear functions are those defined by a linear equation such as $f(x) = ax + b$. Power functions are those defined by a number raised to a power, $f(x) = a^n$. Quadratic functions, just like the name implies, are defined by a quadratic polynomial, $f(x) = ax^2 + bx + c$. Exponential Functions are of two forms, $f(x) = a^x$ to the base $a$, where $a > 0$, and, $f(x) = e^x$ to the base $e$, where $e$ is a number defined by Euler. He also uses the trigonometric functions $f(x) = sin\ x$, and $f(x) = cos\ x$.

$\mathbf{W}$e spent the afternoon at the *Jardin des Tuileries*. Maman and Madeleine were eager to see the latest fashions, as it's strolling through the gardens how women display their newest gowns. I like to go with them because Maman allows me to visit the bookstores nearby. There is one shop I especially like because it offers new and used books at very reasonable prices. I found a new Jean de la Fontaine's *Fables*, with colorful illustrations, much nicer than the old book at home. So I bought it for Angelique.

My study of Euler's book is progressing slowly, as I have to translate the text to understand his analysis, which is a bit challenging. He introduces two mathematical terms: sequence and series. It took several days of study, but now I understand what they mean.

A *sequence* is a set of numbers arranged in an orderly fashion such that the preceding and following numbers are completely specified. The numbers in a sequence are called "terms." The "general term" defines the rule of the sequence. For example, if $n$ is the ordinal number of a term in the sequence, 1 is the first term, and the general term $a_n$ is $2n - 1$, the sequence is written as 1, 3, 5, 7, ..., $2n - 1$.

A *series* is the sum of the terms of a sequence, and it can be a finite or infinite series. Finite sequence and series have defined first and last terms. For example:

$$\sum_{n=0}^{5}\left(\frac{1}{2}\right)^{n} = 1 + \frac{1}{2} + \frac{1}{4} + \frac{1}{8} + \frac{1}{16} + \frac{1}{32}$$

where $n$ takes values from 0 to 5. The symbol $\Sigma$ represents a sum.

Infinite sequences and series continue indefinitely (to infinite). For example:

$$\sum_{n=1}^{\infty}\left(\frac{1}{n}\right) = 1 + \frac{1}{2} + \frac{1}{3} + \frac{1}{4} + ... + \frac{1}{n}$$

where $n$ takes values from 1 to $\infty$.

Euler includes derivations of the power series for the exponential, logarithmic, and trigonometric functions; the factorizations of the sine and cosine functions; and the consequent evaluation of the "sum of the reciprocal squares." Euler defines logarithms as exponents and the trigonometric functions as ratios.

There is a new mathematical expression that Euler used to manipulate the infinite series. First, how do I find the sum of an infinite series? To do it I must find its limit. For example, if I ad more terms to a convergent series, its terms become smaller and smaller. In other words, the terms of a convergent series tend to 0. The sum of such a convergent series is referred to as "the sum to infinity," and mathematicians simply write this as:

$$\lim_{n \to \infty} S_n$$

where *lim* stands for "limit" a term from the Latin word *limes*. The mathematical meaning is still a little confusing, but I'm sure the next sections of Euler's derivation will help me learn what the limit of a function is and how to calculate it.

Friday—April 30, 1790

I continue studying Euler's book. The two volumes of the *Introductio in analysin infinitorum* cover a wide variety of subjects, including the infinite series and expansion of trigonometric functions. Right now I am studying a new number, which Euler calls "*e*." It is a most amazing number that Euler connects to the trigonometric functions in a most intriguing manner. Euler begins by defining the number *e* with an infinite series expansion:

$$e = 1 + \frac{1}{1} + \frac{1}{1 \cdot 2} + \frac{1}{1 \cdot 2 \cdot 3} + \frac{1}{1 \cdot 2 \cdot 3 \cdot 4} + \dots$$

84

Euler shows that the number $e$ is the limit of $(1 + 1/n)^n$ as $n$ tends to infinity. I still do not know the meaning of this number $e$, but I imagine it must be similar to $\pi$. I think so because, by definition, $e$ is equal to an infinite series. This implies that its value is not exact, since one can always add another term to the series and keep on adding to infinity. The digits on the right of the decimal point do not repeat.

Euler gave an approximation for $e$ to 18 decimal places, $e = 2.718281828459045235$, without saying where it came from. Perhaps Euler calculated the value himself, but he did not say how. However, I computed the ratio of the first seven terms in the infinite series to verify the value of $e$ by myself:

$$1 \qquad\qquad = 1.0000000$$

$$\frac{1}{1} \qquad\qquad = 1.0000000$$

$$\frac{1}{1\cdot 2} = \frac{1}{2} \qquad\qquad = 0.5000000$$

$$\frac{1}{1\cdot 2\cdot 3} = \frac{1}{6} \qquad\qquad = 0.1666666$$

$$\frac{1}{1\cdot 2\cdot 3\cdot 4} = \frac{1}{24} \qquad\qquad = 0.0416666$$

$$\frac{1}{1\cdot 2\cdot 3\cdot 4\cdot 5} = \frac{1}{120} = 0.0083333$$

$$\frac{1}{1\cdot 2\cdot 3\cdot 4\cdot 5\cdot 6} = \frac{1}{720} = 0.0013888$$

Adding them up, I get 2.7180553, a number that approaches the value for $e$ given by Euler. So, if I could add all the terms of the series, perhaps I could prove that

$$e = 1 + \frac{1}{1} + \frac{1}{1\cdot 2} + \frac{1}{1\cdot 2\cdot 3} + \frac{1}{1\cdot 2\cdot 3\cdot 4} + \ldots = 2.718228\ldots.$$

Euler also wrote the fraction expansion of $e$, and noted a pattern in the expansion. He did not give a proof that the patterns in

85

the series continue, but I am sure that it should lead to a proof that $e$ is irrational. Whatever the number $e$ means, it has to be irrational, just like $\pi$. And if $\pi$ is related to the circle, what is $e$ related to?

There are other very interesting results in Euler's book. There is a chapter on the "zeta function" (which I still don't know what it means) and its relation to prime numbers, as well as a chapter on *partitio nomerum* (I have not translated this very well). I could go on, but it is late already and I need to sleep.

Monday—May 10, 1790

$I$ wish to share with someone what I am learning from Euler. He derives so many beautiful relations, so simple and yet so intriguing. It's amazing how the mind of one person can create such beauty with numbers. One example of Euler's genius is this equation:

$$e^{i\pi} + 1 = 0$$

It connects four unique fundamental numbers in a relation of exquisite simplicity: the principal whole numbers 1 and 0, the chief mathematical signs $+$ and $=$, and the special numbers, $e$, $i$, and $\pi$. I do not know what the relationship of these numbers means, but I can only imagine the secrets that this beautiful equation guards within.

Euler also shows a connection between my favorite functions: the trigonometric and the complex exponential functions, $e^{ix} = \cos x + i \sin x$. He deduced this relation using De Moivre's formula, which states that for any real number $x$ and any integer $n$, $(\cos x + i \sin x)^n = \cos(nx) + i \sin(nx)$. The relation is important because it connects complex numbers and trigonometry. I do not know what its significance is, so I must attempt to follow Euler's analysis to understand how he developed his equation for $e$. I start with the series expansion of the three functions $e^x$, $\sin(x)$, and $\cos(x)$:

$$e^x = 1 + x + \frac{x^2}{2!} + \frac{x^3}{3!} + ...; \qquad \sin(x) = x - \frac{x^3}{3!} + \frac{x^5}{5!} - ...; \qquad \text{and}$$

$$\cos(x) = 1 - \frac{x^2}{2!} + \frac{x^4}{4!} - ....$$

Then I write:

$$e^{ix} = 1 + i\,x - \frac{x^2}{2!} - i\frac{x^3}{3!} + \frac{x^4}{4!} + ...$$

$$= (1 - \frac{x^2}{2!} + \frac{x^4}{4!} - \frac{x^6}{6!} + ...) + i(x - \frac{x^3}{3!} + \frac{x^5}{5!} - \frac{x^7}{7!} + ...)$$

$$= \cos x + i \sin x$$

The only way to obtain Euler's equation is by substituting the variable, that is, letting $x = \pi$, then I can write:

$$e^{i\pi} = \cos \pi + i \sin \pi = -1 + i \cdot 0 = -1$$

which means that $e^{i\pi} = -1$
or

$$e^{i\pi} + 1 = 0$$

*Voila*! It's beautiful!—but if I didn't know that I could substitute the variable $x$ with the constant number $\pi$, I probably would not have arrived at the expression for $e^{i\pi}$. In his book, Euler did not explain why he made the substitution, or why it is valid. But it works. That's another example of the way mathematicians think. It requires one to know a lot, and to "see" the right connections between equations and numbers, just as Euler did.

Saturday—May 15, 1790

Now I understand this equation: $e^{ix} = \cos (x) + i \sin (x)$. First of all, the equality implies that $e^{ix}$ is a complex number with two components, a real part (*cos x*) and an imaginary part (*sin x*). Furthermore, if $x = \pi$ (or any factor of $\pi$), the value of the exponential function depends on the cyclic value of the trigonometric functions.

For example, with $x = \pi$, $cos\ \pi = -1$ and $sin\ \pi = 0$. Therefore, $e^{i\pi} = cos\ \pi + i\ sin\ \pi = -1 + i{\cdot}0 = -1$, which leads to Euler's relation $e^{i\pi} + 1 = 0$. Euler knew his trigonometry!

Now, if $x = 0$, $cos\ 0 = 1$ and $sin\ 0 = 0$, then I get: $e^{i0} = cos\ 0 + i\ sin\ 0 = 1 + i{\cdot}0 = 1 = e^{0}$. This shows that, indeed, the number $e$ follows the same rules of exponents that I learned in algebra: any number raised to the 0-power is equal to 1.

So, what happens if the imaginary unit $i$ is raised to the $i$-th power? What kind of number $i^i$ is? Is it real or imaginary? I will find out.

Starting with Euler's equation $e^{i\pi} = -1$, which is the same as $e^{i\pi} = i^2$ (since by definition $i = \sqrt{-1}$ or $i^2 = -1$). If I raise both sides of this equation to the $i$-th power, I get,

$$(e^{i\pi})^i = (i^2)^i$$

or

$$e^{i\ i\pi} = e^{-\pi} = (i^2)^i = (i^i)^2$$

Now I take the square root of each side,

$$(e^{-\pi})^{1/2} = [\,(i^i)^2\,]^{1/2}$$

so

$$(e^{-\pi})^{1/2} = i^i$$

which is the same as $\dfrac{1}{(e^{\pi})^{\frac{1}{2}}} = i^i$.

This means that the imaginary power of the imaginary unit is a real number!

What is infinite? The other night, observing the twinkling stars in the sky, I began to wonder about the size of the universe. There are so many stars in the firmament, so far away. Does the cosmos have a limit? Since the universe is made up of matter, it probably has bounds, otherwise there would have to be an infinite amount of matter.

What about the numbers—are they infinite? Yes! Of this I am sure. Because no matter how large a number is, I can always add another one and make it even larger. How large is "large"? If I use an exponent to represent a very large number, say $10^n$ where $n$ is a number greater than any number I could count, I cannot even conceive in my mind the number of digits it would have. Because even if I got tired of counting, I could always add one more number, that is $10^n + 1$, so this number is greater than my original largest number, $10^n + 1 > 10^n$.

Mathematicians define *infinite* as an unbounded quantity greater than every real number. The real numbers include all the integers, rational and irrational numbers. Thus, whatever number I choose to be the last in my mind, I could as easy make it greater by adding 1, and so I could never reach the last number. Furthermore, since both $10^n$ and $10^n + 1$ are already very large numbers, they are infinitely large. I imagine this is like adding a drop of water in the ocean; the drop makes no difference to the size of the ocean. Therefore, adding one to infinity should yield infinite!

On Thursday we celebrated the feast of Corpus Christi at St-Germain church. We went there because Maman wanted to see the king and queen, who lead the procession of the Blessed Sacrament. After all the riots and protests of the past year, the monarchs were gracious to join the people in this manner.

Mother would be mortified and angry with me if she knew that during the Mass, my mind was wandering, distracted by mathematics. I was thinking about two very simple equations that appear to be related; yet each one led me into a different realm of mathematics.

The equations are of the form $x + a = 0$ and $x^2 + a = 0$, where $a$ is any positive number. These two seemingly innocent equations require solutions that are outside the confines of real numbers. The solution of equations such as $x + a = 0$ requires negative numbers. For example, $x + 1 = 0$, has the solution $x = -1$. On the other hand, the solution to $x^2 + a = 0$ is an imaginary number. As I discovered, the number $\sqrt{-1}$, which is called the imaginary unit, is defined as one of the solutions of the quadratic equation $x^2 + 1 = 0$ (the other solution is $-\sqrt{-1}$).

What this means is that the two equations $x + a = 0$ and $x^2 + a = 0$ have no solutions unless we introduce negative and imaginary numbers, respectively. That is why I find mathematics so fascinating. To solve certain equations, even simple ones, sometimes one must develop new mathematics if no solution is possible within the realm of what is already known.

Does this mean that all problems in the universe have solutions? Well, if we identify a problem, indubitably it must have a solution. That's why my dream is to learn mathematics, and devote my life to seeking out answers to difficult questions. My goal is to search for a mathematical concept or solution that hasn't been discovered.

Saturday—June 12, 1790

Ever since I caught sight of a colorful hot-air balloon floating over Paris, I've dreamed of flying among the clouds. That is just a fantasy, of course. Mother would never allow it. But it is wonderful to imagine that I could fly. When I was little, Papa read me the myth of Deadalus and Icarus, who wanted to imitate grace and ease of flying birds. And then I dreamt that I, too, had wings.

I've read about Leonardo da Vinci and his inventions. He studied the flight of birds and sketched flying machines in a world that only he could see. Leonardo wrote: "If man could subjugate the air and rise up into it on large wings of his own making." Oh yes, I know that the human body is not meant to fly like birds, but wouldn't it be wonderful to at least fly aboard a flying machine like the ones in Da Vinci's imagination?

We went to the *Jardin des Tuileries* and then stopped at the Palais Royal. It was crowded and noisy; it seemed as if all Parisians were there today. There is always something new to catch people's attention; it's like going to a festival. We saw acrobats, magicians performing all sorts of astonishing tricks, and vendors selling amusing toys. There were folks dressed in multicolored costumes, dancing and singing in foreign languages. The arched passages were full of ladies wearing beautiful gowns and the most outrageous hats I've ever seen. Mother and Madeleine also visited the *Camp des Tartares*, the part where the shops and boutiques are found.

After shopping, Madeleine met with her fiancé at the Café de Foy. I waited with them while my parents took Angelique to the marionette theater. We sat at a table in the central garden, and M. Lherbette ordered iced lemonade for us. Later Papa took me to a special bookshop that sells scientific books from all over the world.

Papa gave me money to purchase a translation from one of Galileo's greatest works, the one that deals with his astronomical discoveries. The book is titled *Sidereus Nuncius* (*The Starry Messenger*) and describes what he observed with his telescope. The original manuscript was published in Venice in 1610. It is fascinating; Galileo claimed there are mountains on the moon! This is the perfect book to read during the hot nights of summer when the moon shines adorned by the twinkling stars. I must stop writing now to resume reading the *Starry Messenger*.

Thursday—June 17, 1790

It struck me like lightning on a dark night. There it was, the connection that I had not seen until now, the relationship that exists

91

between logarithms and the number $e$. By definition, $e$ is the limit value of the expression $(1 + 1/n)$ raised to the $n$th power, when $n$ increases indefinitely. Seen this way, I did not connect it to logarithms. I had taken logarithms simply as tools to write a very large number in compact form. I knew that if three real numbers $a$, $x$, and $y$ are related as $x = a^y$, then $y$ is defined as the base-$a$ logarithm of $x$. This is also known as the Napier logarithm, after John Napier, who developed it. That is, $log_a x = y$. For example, $1000 = 10^3$, therefore I can write $log_{10} 1000 = 3$. This means that 3 is equal to the logarithm of 1000 when the base is 10.

Then I noted the expression $x = e^y$, and I knew that I could also write: $log_e x = y$, or a logarithm of base equal to $e = 2.71....$ Now I know that to differentiate the logarithm of base 10 from the logarithm of base $e$, mathematicians use the term natural logarithm and use the nomenclature $ln$, such that $log_e x = ln x = y$.

Euler had the mind of a genius, as he discovered such incredible relationships. I am so glad to see the connection, too.

Friday—June 25, 1790

All French citizens are socially equals. Or at least in principle. On Saturday the Assembly voted to abolish the nobility, including the removal of aristocratic titles. That means that Monsieur the Marquis de Condorcet will now be addressed simply as Monsieur Condorcet, and the Duchess de Quelen will be, from now on, just Madame de Quelen. But removing their title—does it make aristocrats equal to the peasants in the countryside, or the workers in the factory? I do not think so. I don't believe that it is possible to achieve true social equality. Some people will always have more wealth and power than others; how can the poor become equals with them?

However, who needs aristocratic titles when one is an educated person? I am more impressed by scholars, by people who are recognized by their intelligence and their erudition. For sure, the Marquis de Condorcet would rather be called "monsieur the mathematician" than "monsieur the marquis." I believe that.

I read a biographical sketch of the great mathematician Leonhard Euler. He did not have an aristocratic title, but he achieved immortality! He was born on April 15, 1707 in Basel, Switzerland. When he was fourteen, he entered the University of Basel; three years later, he was granted a master's degree in philosophy. His father expected him to become a minister, but Euler liked mathematics better. His teacher, Johann Bernoulli, who was a famous mathematician, convinced Euler's father to let him study mathematics. And that is how Euler became a mathematician. In 1730, he became professor of physics at the Academy of Sciences in St. Petersburg, and three years later he became professor of mathematics. Euler married the daughter of a Swiss painter and had thirteen children.

Euler had an extraordinary memory. I read that once Euler did a calculation in his head to settle an argument between students whose computations differed in the fiftieth decimal place. Euler was a very prolific mathematical writer and published hundreds of scientific papers. He won the Paris Academy Prize twelve times! Not everything was fine with him. Euler lost sight in his right eye when he was relatively young and became completely blind many years later. Nevertheless, aided by his amazing memory, and having practiced writing on a large slate when his sight was failing him, he continued to publish his results by dictating them to his pupils. I don't know of any other mathematician who achieved so much.

Leonhard Euler not only made advancements in mathematics, but also in astronomy, mechanics, optics, and acoustics. Euler was also first to prove that $e$ is irrational. I could go on and on writing about Euler's many achievements. This inspires me to learn more mathematics and attempt to become a little bit like him.

Tuesday—July 6, 1790

Are women equal to men? Are we now allowed to attend the university like boys do? Well, according to Papa, today is the beginning of a something that will benefit women. I'm so euphoric

93

that I am getting ahead of myself. Let's start from the beginning. I had not planed to be in Papa's study this evening when his friends came to their regular Tuesday meeting. But they arrived early, and Papa asked me to stay. I didn't know why it was important since I never participate in their discussions. Soon it became clear. When all of his friends were comfortably seated, my father pulled out of his desk a pamphlet that he waved in his hand. "This," Papa said excitedly, "this is what France needs—complete civil equality for women and men!"

Papa explained that the Marquis de Condorcet had just published an essay arguing in favor of granting the same rights of citizenship to women. I was stunned. The gentleman who had taken seriously my interest and desire to learn mathematics is indeed the enlightened man I had assumed he would be.

Papa's friends, however, had different points of view on the matter. Some argued that "women have never been guided by reason and, therefore, they could not possibly use good judgment" in matters related to political and civil duties. Others even claimed that men have superiority of mind. However, my father refuted those silly arguments, and the discussion became very heated. At that moment, Papa placed M. Condorcet's pamphlet in my hand, and I dashed off with it. I came to my room to read it.

M. Condorcet's essay is titled *On the Admission of Women to the Rights of Citizenship*. In it he writes that "either no member of the human race has real rights, or else all have the same; he who votes against the rights of another, whatever his religion, colour, or sex, thereby abjures his own." The most interesting part of M. Condorcet's essay is when he talks about education for women. He states that "education must be the same for women as for men."

Condorcet argues that both men and women should pursue learning in common, not separately. He states it is absurd to exclude women from training for professions, which should be open to both sexes on a competitive basis. He even says that women should have equal opportunities to teach at all levels; "because of their special aptitudes for certain practical arts, theoretical scientific studies would be of particular value to them."

This is the first time someone advocates publicly education for women. It is certainly a new way of thinking in sharp contrast with the old-fashioned beliefs of most people in France.

Tomorrow is the first anniversary of the republic. My sisters are preparing the dresses they'll wear at the festival. Everybody is excited, except Mother, who is still upset by the memory of the massacres during the assault on the Bastille.

The *Fête de la Federation*, Festival of the Federation, will take place on the fields of the Champ de Mars near the military school. Mother's only justification for going is to see Queen Marie Antoinette and the royal family who will preside at the festival. Madeleine and I are going to wear new white dresses and straw hats with red, white, and blue ribbons. Angelique wants to wear her long hair down, adorned with a garland of flowers decorated with flowing tricolor streamers that she designed herself. She is so proud of her creation and can't wait until tomorrow to show it off.

Many people from the provinces have arrived in Paris already. The cafés are crowded, the streets are full of visitors; the city is bustling with anticipation. For three weeks many people have worked to prepare the Champ de Mars for the festival. Millie said that even women and children were helping the workers.

Father and his friends gathered to discuss the events of yesterday at the meeting of the National Assembly. The delegates passed laws purporting to make all members of the clergy subject of the State. What this means is that, from now on, the priests will be paid by the government and will be elected by the State or district electors. The priests will have to swear an oath of allegiance to the nation, and the government will regulate their conduct. Any priest who refuses to swear the oath will be forbidden to perform any of his duties and will be subject to arrest and punishment if he does. The priests who swear the required oath will be called government clergy, and those who refuse will be called refractory clergy.

Mother is troubled by this turn of events. She argued that the pope is the head of the Catholic Church, not the Assembly, and that the government should not rule the clergy. Father simply noted that the new Civil Constitution of the Clergy is official and nothing can be done about it. He tried to remind her that, since February, monastic vows were forbidden, so the clergy must be elected by the State, or there won't be priests at all.

The church as we knew it no longer exists. For sure this is going to affect the way people worship. My mother claims that if the priests take the oath, she will not take communion from them. We'll see.

Wednesday—July 14, 1790

We celebrated the *Fête de la Federation* in the rain. We awoke early at the sound of thunder. However, not even rain could dampen my sister's enthusiasm for the festivities. After morning prayers, Angelique ran around impatiently, urging everybody to hurry and finish breakfast. She did not want to miss the parade that preceded the festival. Angelique hardly ate, prodding Millie to help her get dressed.

At midmorning, my sister called us excitedly when the first sounds of the military drums were heard in the distance. The parade started at the Temple, advancing along the rue Saint-Denis. Marching to the sound of artillery salvos and military bands were soldiers, the National Guard, government delegates, sailors, and even a battalion of children bearing a banner declaring they were "the hope of the *patrie.*" People in the streets dropped flowers as they passed by. My sister and Millie threw kisses and waved at the soldiers who marched proudly wearing wet uniforms and squelching boots.

After the procession turned westward onto the rue Saint-Honoré, we followed it by carriage. We dropped Papa at the Place Louis XV, the plaza where he reunited with the deputies of the National Assembly to join the procession. We kept riding the carriage on our way to the Champ de Mars. It was already one in the afternoon when we arrived, entering the amphitheater through the triumphal arch full of people. A huge crowd—people from every class and all stations of society, workers, bourgeois families, peasants, and aristocrats—waited for the celebration to begin. Music was playing, making everybody merry and gay; many were singing patriotic songs and dancing in the rain.

When King Louis and his court arrived, he sat on a blue velvet throne on a high platform. At his side was Queen Marie Antoinette, wearing a beautiful red, white, and blue cape with feathers. We were seated not too far from the platform so we could see their faces clearly. There was also a handsome man on horseback, smartly dressed with military insignias. Mother pointed to the man on the white horse to tell us he was the Marquis de La Fayette, the commander of the *Garde Nationale*.

At three thirty, hundreds of priests wearing tricolor sashes under their white albs made an entrance. A *Te Deum* was sung at the special Mass, a hymn written by Joseph Gossec to celebrate national unity. Gossec set a traditional Latin text to music scored for wind instruments, the sound of which made people tremble with emotion. At the end, the bishop of Autun, Charles Talleyrand, blessed the crowd. The Marquis de La Fayette advanced, posed his sword on the burning flame, and pronounced an oath of fidelity to the nation. All the delegates repeated the oath.

The queen lifted the Dauphin on her arms and people cheered. They responded to the sight of the beautiful child with a demonstration of love for the royal family. Everybody rejoiced in a grand display of unity and brotherhood, waving flags and colorful banners. After the crowds quieted down, the king rose on the platform and pledged to see the laws upheld, and to support the new Constitution of France. After hearing these words, the crowd shouted enthusiastically "*Vive le roi, vive le reine, vive le dauphin!*" "Long live the king, long live the queen, long live the dauphin!"

On this day the people of France—rich and poor, young and old—were in harmony with the monarchy, with the government, and with one another. I hope this marks the end of the chaos and the social conflicts of the past year.

Sunday—July 18, 1790

The *Fête de la Federation* continues. This afternoon we saw a splendid water festival on the Siene River, with music, jousting, and dancing. It is now almost midnight, and the party on the riverbanks is still going on. I can hear the sounds of music and the laughter of people. Angelique, Millie, and I sat by the loft windows this evening and watched the fireworks. The silly girls squealed with delight at every whooshing rocket shooting up, and covered their ears at the powerful blasts. The final bouquet of glittering exploding powder was the most spectacular, as it lit the sky with many breathtaking colors.

My parents and sisters have gone to bed. I, however, feel restless and remain in my reverie. It is not that I am disturbed by the clamor of the crowd celebrating near the river. The truth is, when I concentrate on my studies, I lose sight of my surroundings. But tonight I feel agitated, anxious to write down a revelation that came to me unexpectedly.

Some time ago I came across a mathematical relationship that intrigued me, but I did not know its significance until now. When I calculated the compound interest for my father, I used the relationship $A = P (1 + R/n)^{nt}$. Then I concluded that, as the value of $n$ increases, the term $(1 + R/n)^n$ is bounded.

It just occurred to me that the expression indeed has a limit and the limit is $e$, the number discovered by the great mathematician Leonhard Euler. Looking back at my notes, I wrote that $e$ is the limit of $(1 + 1/n)^n$. This expression is exactly like the term in the formula for the compound interest, with $R = 1$. Now I understand why the total amount paid ($A$) did not increase beyond 2.715 times, even when I compounded the interest every minute. I must tell Papa that there is a well defined mathematical limit to the computation, so he might as well settle for a reasonable number of times a year to compound the interest on his loans.

*Très bien.* So, if $e$ is the limit of $(1 + 1/n)^n$, then I can write, for very large value of $n$ ($n$ approaching infinity, i.e., $n \to \infty$): $A = P e^t$. This expression means that $A$ grows exponentially, and that, even if interest is compounded continuously, the return on a loan is *finite* since the limit of the compound function is $e$. What a neat way to represent continuous compounding.

I once wondered what $e$ means. Knowing that $\pi$ represents the ratio of the circumference to the diameter of the circle, what about $e$? Does it have a physical connection? At least for the problem of compounding interest, I'd say that the number $e$ represents the limit to greed! Seriously, $e$ seems to represent the limit to growth, which makes sense since otherwise things would expand and grow continuously, without bound, to infinite!

I must sleep now to get up early and be ready for the long trip to Lisieux.


Sunday—September 5, 1790

I waited until now to write, held back by a promise I made to my mother. When we arrived in Lisieux, Mother took away my pencils to discourage my analytical work. At first I resented her, but then I realized that probably it was good idea to let my mind rest for a while. However, during our holiday I read philosophy and history books. In the evenings I read poems and novels to Mother. She was bedridden with some strange ailment for a couple of weeks. Papa stayed behind in Paris, but a week later he went to join us. Papa and I played chess and had a few conversations about philosophy and science.

Angelique and I spent many splendid afternoons watching the peasants work the lands. We saw young girls milking the cows and their mothers making cider. Some times, the women in the village invited us to help them wrap the cheese they make for the market. Oh, but the best time we had was playing with kites. I taught Angelique how to make diamond-shaped kites using light frames of wood covered with colorful tissue paper. We spent hours flying the kites, watching the streamers undulating vigorously by the strong wind. We ran across the fields to fly the kites higher, laughing when the kites fell down and holding our breath when they flew so far away that we lost our kites to the wind.

Many nights I spent in solitude, watching the astonishing night sky studded with millions of stars shinning like diamonds. I wondered about the moon growing and disappearing from the sky. I tracked its progress from a huge luminous disk on July 26 to a thin silver crescent on August 8. The cycle repeated and once again the moon was full two weeks later. We left Lisieux on the first of September, and that night I saw the moon waning.

I am glad for the holiday in the country, and to have the opportunity to witness the splendor of the heavens. Now I am ready to resume my studies, anxious to seek new mathematical challenges.

Friday—September 10, 1790

I began to study infinite series. Everything was going well until I came across the following:

$$\frac{1}{(x + 1)^2} = 1 - 2x + 3x^2 - 4x^3 + \ldots$$

When I substituted $x = -1$, it yielded $\dfrac{1}{(x + 1)^2} = \infty$ (since one over zero is infinite).

So, for $x = -1$

$$\infty = 1 + 2 + 3 + 4 + \ldots$$

This result makes sense; it implies that adding all the integers in the universe would result in an infinite value.

Now, there is another series: $\dfrac{1}{1 - x} = 1 + x + x^2 + x^3 + \ldots$

When I substitute $x = 2$, the left hand side of the series is equal to $-1$, that is,

$$-1 = 1 + 2 + 4 + 8 + \ldots$$

Comparing the two series I notice that the second series (for $-1$) is term by term greater than the first series (for $\infty$); in other words, I get,

$$-1 > \infty$$

How is this possible? How can minus one be greater than infinity?

Now, if $x = -1$, the series yields $\dfrac{1}{1 - x} = \dfrac{1}{2}$

Therefore,

100

$$\tfrac{1}{2} = 1 - 1 + 1 - 1 + \dots$$

This does not seem logical either.

Unsure about my analysis, I decided to do more research. I discovered that Euler referred to such seemingly absurd result as "paradox." A paradox, according to the dictionary, is "a statement that seems contradictory, unbelievable, or absurd but that may be true in fact." So, perhaps my two contradictory results are mathematically correct; therefore, I should call them paradoxes. If so, why are there paradoxes in mathematics?

Wednesday—September 15, 1790

Mathematical paradoxes have intrigued philosophers since ancient times. Allegedly, the poet Epimenides once said back in the sixth century: "All Cretans are liars." The statement is called *the Cretan Paradox* because it was uttered by a Cretan, thus is true if and only if it is false. This is the earliest known attempt at formulating a mathematical paradox.

However, it seems that the statement of Epimenides, "All Cretans are liars," is not paradoxical but merely contingent. For example, suppose there are 100 Cretans in the world, one of them being Epimenides. By saying "All Cretans are liars," Epimenides implies that all 100 Cretans are liars; in other words, that all 100 Cretans that exist in the world always make false statements. Could Epimenides be a liar? Yes, simply if one of the other Cretans doesn't always lie. Then Epimendes' statement is simply a false statement, as one would expect from a liar.

Epimenides can't be a truth-teller, because Cretans can only say the truth. "All Cretans are liars" implies "Epimenides is a liar," so if "All Cretans are liars" is true, then so is "Epimenides is a liar," which is a contradiction, hence Epimenides cannot be a truth-teller.

However, if Epimenides was the only Cretan, then "All Cretans are liars" is a paradox, since the statement implies that he is not telling the truth, or that Epimenides is saying, "I am a liar." It is a little confusing; that is why paradoxes are studied in logic.

Now, returning to infinite series, I think I understand why my result seemed paradoxical. Euler wrote about it: "Notable enough, however, are the controversies over the series $1 - 1 + 1 - 1 + 1 - \dots$ whose sum was given by Leibniz as ½, although others disagree. ... Understanding of this question is to be sought in the word 'sum;' this idea, if thus conceived— namely, the sum of a series is said to be that quantity to which it is brought closer as more terms of the series are taken—has relevance only for convergent series, and we should in general give up the idea of sum for divergent series."

Euler was referring to Gottfried Wilhelm Leibniz, a German philosopher, mathematician, and logician who died when Euler was nine years old. Thus, if Euler and other mathematicians deal with this sort of paradox, I should also understand how to separate a paradox from an untrue mathematical statement. This I'll need to evaluate results from analysis that may appear contradictory.

Monday—September 20, 1790

$P$rime numbers are fascinating! Of all the numbers in mathematics, primes are my favorite. Of course I like all numbers, but the primes are more interesting. I am not the only one attracted to prime numbers. Many ancient mathematicians considered them mystical numbers. Pythagoras and his followers, for example, believed that prime numbers had spiritual properties. They liked primes because they are pure and simple, not like the square root of 2 that cannot be expressed exactly as the ratio of two whole numbers.

There was a Prussian mathematician and historian who devoted a great deal of study to prime numbers. His name was Christian Goldbach. He corresponded with Leonhard Euler describing some of his ideas related to prime numbers. In a letter, Goldbach theorized: "Every even number can be written as a sum of two odd primes." This statement is known as the *Goldbach conjecture* because he inferred his statement from uncertain evidence.

I can verify Goldbach's conjecture for any three even numbers chosen arbitrarily, say: 8, 20, and 42.

$8 = 3 + 5$.
$20 = 13 + 7 = 17 + 3$.
$42 = 23 + 19 = 29 + 13 = 31 + 11 = 37 + 5$.

*Eh bien,* I've written each even number as the sum of two numbers. Now I need to check if 3, 5, 17, and 37 are primes. Using the definition, I derive the first primes. *A prime number is a positive integer p > 1 that has no positive integer divisors other than 1 and p itself.* As I determined before, the first primes are: 2, 3, 5, 7, 11, 13, 17, 19, 23, 29, 31, 37, 41, 43, 47, 53, 59, 61, 67, 71, 73, 79, ....

Yes, I can see that 3, 5, 17, and 37 are primes. Indeed, even numbers can be written as a sum of two odd prime numbers, confirming Goldbach's conjecture, at least for the even numbers I chose at random. Furthermore, I notice that there is more than one Goldbach pair (sums of two odd primes), as the number gets larger. Is this to be expected? Goldbach's conjecture says only that there is at least one sum, and it does not to say whether there may be more pairs. But it just seems logical to me that there should be more pairs as the number is greater because there are many more primes available to make up the sums.

Goldbach also predicted that "every sufficiently large integer can be written as the sum of three primes." When he said "sufficiently large," Goldbach must have meant an integer greater or equal to 6 because it does not work with smaller integers. For 6 and beyond, I confirm Goldbach's conjecture:

$6 = 2 + 2 + 2$
$7 = 2 + 2 + 3$
$8 = 2 + 3 + 3$

and so on.

It looks quite simple. However, nobody has proved Goldbach's conjecture for all primes.

$T$his afternoon Angelique and I rode with my parents to the Palais Royal. While Papa met with his colleagues at the Café de Foy, Maman took us shopping at a fashionable boutique. She bought a beautiful fan adorned with pictures of the royal family, painted in bright lovely colors. Later Angelique sat for a silhouette portrait at the studio next to the café. Maman wanted me to have my silhouette made as well, but I refused. I do not like portraits of any kind. I do not like my looks. I don't even like to see myself in a mirror. My sister asked with a scoff, "What do you like, Sophie?" I wanted to scream, but didn't, "I like mathematics!"

I've resumed my studies of infinite series. This topic is full of unexpected results. Just this evening I learned that the number $\pi$ appears in many other equations that are not related to the circle. Euler discovered connections between $\pi$ and infinite series. Euler showed some rather amazing infinite series whose limits involve $\pi$, such as these:

$$\sum_{n=1}^{\infty} \frac{1}{(2n-1)^2} = \frac{1}{1^2} + \frac{1}{3^2} + \frac{1}{5^2} + \frac{1}{7^2} + \ldots = \frac{\pi^2}{8}$$

$$\sum_{n=1}^{\infty} \frac{(-1)^{n+1}}{n^2} = \frac{1}{1^2} - \frac{1}{2^2} + \frac{1}{3^2} - \frac{1}{4^2} + \ldots = \frac{\pi^2}{12}$$

I find it astonishing that $\pi$ would appear as the sum of infinite series! Seeing $\pi$ related to a concept that represents infinity makes one realize that $\pi$ is indeed a number whose exact value will never be known, making it a truly magical number.

How did Euler see the connection between infinite series and $\pi$? I've studied Euler's books for months, and I still cannot grasp everything in them. There is so much beauty in his analysis, but Euler does not tell me what kind of revelations helped him discover such splendor.

I met a most unusual girl who loves mathematics as much as I do. I met her at Madame Geoffroy's party last night. At first I didn't want to go but now I'm glad, as I learned something interesting to include in my studies. But I am getting ahead of myself. First, I want to record what happened last night and how I ended up chatting with a Spanish girl who likes mathematics as much as I.

It started with the invitation to attend a musical soirée at the home of Mme Geoffroy. Maman did not give me a choice, so I dragged myself along and rode with her and my sisters. After the concert they left me alone while they socialized. As always, I felt awkward and uncomfortable. Sitting alone in a corner, I tried to stay away from the idle gossip of the women. Madame Geoffroy approached to present me to a girl. I had seen her earlier, with her black hair and dark eyes contrasting with her modest blue gown. Her name is Maria Josefa Ibarra de la Villa Real.

After Mme Geoffroy left us, the girl told me she is fifteen, and like me, she is fascinated by mathematics. Maria Josefa complained that her family does not appreciate her talents or support her studies. She told me that, since she was five years old, she could perform calculations in her head and could see patterns in numbers. By the time she was ten, Josefa said, she had mastered algebra and geometry, teaching herself from books in her home library, just like me!

Josefa is from a noble family from Aragon in Spain, but after her father died last year, her mother discovered that the family's wealth was almost exhausted. Thus, to save the family's honor and position in society, the mother arranged an engagement for Josefa to marry a rich duke. The poor girl is terrified but feels that she has no choice but to obey. The wedding was to take place in August but, to gain some time, Josefa proposed a trip to Paris to learn the French language and customs, since her fiancé is from here. I noted that she spoke perfect French, so Josefa admitted this trip was just an excuse to postpone the wedding.

She also confided her desire to meet the mathematicians at the Academy of Sciences in Paris. Her dream is to learn from them before she goes back to Spain, although her mother, who accompanies her, may not permit it.

Josefa read many of the books I've studied. She asked me a question about geometry and told me about an intriguing problem she solved, referring to it as the *Delian* problem. I did not have time to ask her about it because Maman went to fetch me; it was time to leave.

I'm glad I met this girl who's clearly very smart and is probably ahead of me in her studies of mathematics. And like me she struggles because her talent is not recognized by anyone. She fears that once she's married, the desires of her heart and her intellect may be drowned by her new responsibilities.

Would my parents also force me to get married? I hope not. Unlike my sister Madeleine, who was anxious to wed, marriage is not a suitable option for me.

Friday—October 8, 1790

Yesterday, while riding the carriage in our way to visit Mme LeBlanc, we encountered a group of men at a street corner shouting political slogans and distributing pamphlets. The men, dressed in long pants without stockings, are members of the revolutionary movement. Because they wear the trousers of the workingman, Papa referred to them as *sans-culottes*. By dressing this way the sans-culottes thumb their noses at the aristocrats.

The conversation at M. LeBlanc's gathering centered on the strong campaigns against the monarchy. The men talked about the political movement that is sweeping the new government. M. LeBlanc claimed that the most aggressive sans-culottes are a menace, shouting and complaining against the monarchy during the meetings of the Assembly. They fight with anyone who is against them. Many get drunk and disturb people in the streets with their speeches against the king. The sans-culottes carry their sabers and claim that they will cut off the ears of all enemies of the nation, and, at the first sound of the drum, they are ready to fight.

This morning Millie complained to my mother that a group of sans-culottes was taunting the women waiting in the long lines at the bakery. They made speeches inciting them to march against King Louis, and they promised to kill all the aristocrats and rich abbots.

Papa contends that not all sans-culottes are bad people. The leaders of the revolutionary movement include lawyers who made significant contributions to the writings in the new Constitution. However, he admitted that some of them irrationally seek revenge for social inequalities and real or perceived injustices committed in the past by aristocrats. There are always people who lean to the extreme of any ideology. Many sans-culottes are political extremists, and their actions might hinder, rather than help, the ideals of the new French Republic.

Wednesday—October 20, 1790

Monsieur LeBlanc talked briefly with me about mathematics. He told me Antoine is studying trigonometry because he wants to be an engineer. M. LeBlanc asked me whether I knew the trigonometric functions. Of course I know trigonometric functions, but he made me doubt my knowledge.

Right after dinner, I went to review my notes and found that I had written the definition: "Trigonometry is the study of angles and of the angular relationships of planar and three-dimensional figures." The word "trigonometry" is derived from the Greek *trigônos,* for triangle, and *metron,* for measure. The trigonometric functions, also called the circular functions, comprising trigonometry are the sine, cosine, tangent, etc., written in mathematical equations as *sin x, cos x, tan x,* and so forth. The trigonometric functions are related to each other. I encountered the trigonometric functions in Euler's book, where I discovered how they are interrelated, not only with themselves, but also with the exponential and the logarithmic functions.

Without realizing it, I began to study trigonometry the moment I was introduced to the Pythagorean Theorem. Due to the nature of the mathematical relation, $z^2 = x^2 + y^2$, at that time I

107

focused mainly on learning to calculate the hypotenuse of right triangles. I remember learning how an angle, in a right triangle, could be calculated by either determining its sine or cosine, depending on whether I knew the opposite side, the adjacent side, plus the hypotenuse, or determining the tangent of the angle if I knew the opposite and adjacent sides of the triangle.

After reviewing my notes, I came up with another startling revelation. Trigonometry is also interrelated with imaginary numbers. From Euler I learned that the number $e$ raised to an imaginary power is equal to a number that has two parts, a real and an imaginary part, which happens to be the trigonometric functions sine and cosine:

$$e^{i\alpha} = \cos\alpha + i\sin\alpha$$

where $\alpha$ is any given angle.

In the same manner, for a negative power I can write

$$e^{-i\alpha} = \cos(-\alpha) + i\sin(-\alpha)$$
$$= \cos\alpha - i\sin\alpha$$

so that

$$e^{i\alpha} \cdot e^{-i\alpha} = (\cos\alpha + i\sin\alpha)(\cos\alpha - i\sin\alpha)$$
$$= \cos^2\alpha - i^2\sin^2\alpha$$
$$= \cos^2\alpha + \sin^2\alpha = 1$$

where I used the fact that $i^2 = -1$

Thus, $e^{-i\alpha} = \dfrac{1}{e^{i\alpha}}$

*Voila!*

$W$hat do I get if I raise Euler's number $e$ to a negative power? Exponential functions such as $e^x$, where $x$ is a positive number, represent growth, just like in the case of the compounded interest. I can also write $e^{-x}$, since, by the rule of exponents, $e^{-x} = 1/e^x$. In this case, $e^{-x}$ represents the opposite of growth, which I call "decay."

Also, the natural logarithm $ln$ is the inverse of Euler's number $e$. Since $x = e^y$, taking the natural log of both sides I obtain $ln\ x = y$. If I have a function $A = C\ e^{ky}$, where $C$ and $k$ are any positive constant numbers, then I rewrite the exponential function as $A/C = e^{ky}$, and taking the natural logarithm of each side of this expression I get

$$ln\ (A/C) = ln\ (e^{ky}) = ky$$

Thus, to solve for $y$, I simply take the natural logarithm of the ratio $A/C$ and divide by the constant $k$. This is too easy. I should try another problem.

For example, the equation: $ln\ (y - 1) + ln\ (y + 1) = x$, where $y$ represents the unknown.

Using the rule of logarithms, $ln\ a + ln\ b = ln\ (a{\cdot}b)$, I can rewrite the equation as:

$$ln\ [(y - 1)\ (y + 1)] = x$$

And since $(y - 1)(y + 1) = y^2 - 1$

$$ln\ (y^2 - 1) = x$$

To solve for $y$, I recall that a natural logarithm is the inverse of $e$; thus, the terms on both sides of the equation are raised to the $e$ power as follows:

$$e^{[\ln(y^2 - 1)]} = e^x$$

And since $e^{\ln a} = a$, the expression reduces to

$$y^2 - 1 = e^x$$

or $\qquad y^2 = 1 + e^x$

And from this expression I can solve for $y$. And the analysis is done.

Saturday—November 27, 1790

The deputies of the National Assembly are going after the dissenting bishops and other powerful clergy who are against their authority. The Assembly issued the Civil Constitution of the Clergy that requires that every priest swear an oath of loyalty to the State. The oath is meant as a legal means to rid France of the corrupted heads of the Church. If a priest does not swear the oath, he will not be allowed to perform the sacraments.

My mother sees this turn of events as blasphemy because she believes the clergy owe obedience only to the Holy Church. I don't know what to believe. What is troublesome is that many citizens are taking their frustrations against innocent priests. Some misguided people are breaking in convents and desecrating churches. I don't think this is the way to eradicate the corrupted church leaders. The Assembly should leave alone the parish priests who are innocent and harmless to the republic.

Saturday—December 25, 1790

It is terribly cold tonight. The ink is freezing in my pen, making my writing illegible. But I have to record what happened to me this evening. I feel like crying and I wish to disappear into a

110

place where nobody can find me. Why can't I be like my sisters? They are pretty, extroverted, and everybody likes them. I, on the other hand, do not know how to behave and end up making a fool of myself in front of people. We had guests this evening. Madeleine and her husband, Monsieur Lherbette, M. and Mme de Maillard, and M. and Mme LeBlanc and their son Antoine-August came for dinner. I tried to be sophisticated and mature, but I ended up looking like a little girl reciting a foolish tale. Oh, I am so mortified!

It started at dinner. I had not seen Antoine-August for a long time. I could not recognize him, as he's become quite snobbish. At fifteen, he has adopted an air of superiority and speaks so pedantically. He was wearing a lot of *eau de cologne* and powdered curls that made him look like a silly old man. He hardly smiled, trying to appear as a serious intellectual. He barely spoke to me throughout dinner.

After the meal we went into the living room. Everybody was in a cheery mood, laughing and having a good time, but I felt awkward and out of place. The party started when Angelique and Madeleine played a lovely sonata. Maman followed, singing a soulful aria, and then Papa and M. LeBlanc took turns reciting poems from Racine's plays. I was nervous, knowing that my turn would come and someone would ask me to play the piano.

I don't like to play in front of people because I am not as good as my sisters. So, when Maman turned to me, I asked if I could tell them an interesting story. Papa thought it was a great idea, so I chose to tell them about Dido, the queen and founder of Carthage. Antoine, who acts as if he knows everything, asked if this was going to be a long story (he meant boring) and almost made me desist. My father came to me and, holding my hand, invited everybody to listen. I decided to tell the story of the Phoenician princess Dido to avoid playing the piano.

Dido had to flee from her homeland because her tyrant brother, Pygmalion, had killed her father and was after her wealth. Dido and her loyal followers sailed away until they reached the coast of Africa. Dido wished to find a place to build a city and start all over. When Dido and her people arrived to a place in the coast of Africa, she asked the ruler to sell her a piece of land. For the money Dido offered, the king promised her as much land as might lie within the boundaries of a bull's hide. Any other person would have given up, thinking that the piece of land would be too small, but Dido was a clever woman. She had her slaves cut the hide into many thin strips and tie them together to form a very long one. She then ordered them to lay the long strip over the land, except over the seashore facing the Mediterranean Sea. Dido wanted to secure as extensive a territory as possible within this boundary, so her slaves laid the skin strip into a half-circle over the land. Choosing this shape, Dido got a much bigger piece of land than the king had thought possible. On this land Dido founded the great city of Carthage where she ruled for many years.

There was silence after I finished talking. I am not sure if they understood the significance of Dido's story. I asked them, "How did Dido know that in order to get more territory next to the seashore she had to choose a piece of land in the form of a semicircle?" Yawning, Antoine-August replied that everybody knows that the circle encloses the maximum area. To that my father replied, "But do you know why?" When there was no answer, Papa said proudly, "Well, my Sophie, *ma petite élève,* she certainly knows!"

Antoine-August wanted to show me that I was wrong by half. So, before he did, I explained why Dido placed her strips of bullhide in a semicircle and not in a circle. Since she was using the shoreline as a boundary for her land, she did not need to waste the strip for that part of the boundary. Under that condition, the semicircle gave her indeed the largest area. Of course, if the clever princess would have to choose a piece of land away from the seacoast, for sure she would have selected it in the shape of a circle.

After that uncomfortable moment the party became unbearable for me. No one made any comments but after an awkward silence, everybody resumed their singing. I felt uneasy but nobody noticed; the party continued without me. And even after the guests left I couldn't stop feeling like a fool.

The church bells tolled the midnight hour long ago, and I am still awake, rehashing the incident. But after pouring my soul in this writing, I do not feel like crying anymore.

Friday—December 31, 1790

It's past midnight, and Paris life is in full swing. I hear the laughter, the singing from merry folks on their way to balls and dances. Some people amble through dark alleys, seeking the cheery lighted avenues; others ride in their cozy carriages. Today people rejoice; they do not care about affairs of state or economic troubles. On this celebrating night there are no political speeches, no calls to arms.

King Louis XVI succumbed to the pressure of the National Assembly and granted his public assent to the Civil Constitution of the Clergy. Now the radical reformers have royal sanction to proceed with their plan for submission of the clergy. The revolutionary government will impose formal and public oaths of allegiance from the priests. At dinner, my parents discussed it, saying that the clergy has one more day to decide whether or not to take the oath. Those who refuse to take the Oath of the Clergy will be removed and replaced by more malleable favorites, or the Assembly will create new priests. It's absurd. And yet, this is the new political reality in my country; the new government is taking control of the Church of France.

My parents say that I am too young to understand politics. Maybe that is why I cannot comprehend what is the fundamental purpose of the revolutionary movement in France. As Papa said tonight, one day it will become clear. But for now my mind must be devoted to the pursuit of my studies.

One year ends and another begins. Just like numbers, one after another in a long infinite sequence, time progresses instant by instant. And life follows its course ...

CZ8O

*Paris, France*
*Year 1791*

# 3
## Introspection

THE BELLS FOR SUNDAY MASS tolled as usual this morning. However, today was no ordinary Sunday. The Catholic ritual at Notre Dame was transformed into a civil ceremony staged by the government. Upon arrival, I knew something different was about to happen. The church was full with army officers dressed in magnificent tricolor uniforms, and instead of sacred music, an orchestra played a patriotic march. This was the day when many priests took an oath to uphold and obey France's new constitution.

The ceremony began when the officers marched to the altar followed by the priests and bishops. At the altar, the clergy pledged to be loyal to the State. Some people cheered when the priests took the oath, but many more booed. Mother didn't express her disappointment until we came back home. It is difficult for her to see the priests renege their vows to the Holy Church. Papa tried to justify the need for the oath, saying that the Assembly has no intention of tampering with the people's beliefs or Christian ceremonies. The new Constitution is only intended to create a national French Church devoid of corruption. Mother was not appeased. She fears the rituals will be less about God and more about politics.

Thursday—January 6, 1791

The Feast of the Epiphany is one of Angelique's favorite celebrations. She likes it because she receives presents to commemorate the gifts of the Magi. After morning prayers, Madame

Morrell set the table with hot chocolate and a special cake. Angelique had to wait until Madeleine and her husband came for supper to get her gift. They gave her a set of charcoal pencils and a drawing pad. I should not get gifts because I am not a little girl anymore, but Madeleine gave me a set of dominoes made of stone and marble that was supposedly torn from the ruins of the Bastille. I don't play dominoes but I thanked M. Lherbette anyway, as probably he chose it. Maman gave Angelique a little mirror encased in white porcelain, and a comb made of amber. Papa gave her a puppet dressed as a court jester.

M. and Mme Geoffroy came later in the evening. The jovial mood of the party turned into a heated discussion when other friends arrived. The women went into Mother's sitting room to gossip, and I took Angelique to my room to read. Later Millie joined us and made us laugh, telling us some funny stories from a play. Millie loves the *Comédie Italienne*. She says that's the best entertainment in Paris. A little silly and sometimes too loud, but Millie is quite smart and would benefit if one day society changes to allow women a formal education. Right now, however, women are not allowed in the universities.

But there is hope that the future will be better for all the women of France.

Thursday—January 20, 1791

It is bitterly cold. The snow keeps falling, piling on the streets like a thick white blanket covering the ground. The city is silent, so quiet as if devoid of people. The only sound I heard this morning was the murmur from my mother praying. She prayed for King Louis who is ill; it is rumored that he has fever and is coughing up blood. Maman says those are symptoms of tuberculosis.

I advance in my studies. The more I learn, the greater my fascination with numbers. It's amazing how many things one can do with numbers. I like algebra in particular; I can spend hours and hours manipulating equations. Initially I had the impression that geometry involved only calculation of areas and angles, but I changed my mind. I learned that numbers and equations define all geometric shapes. After meeting the Spanish girl, I came to learn how the Greeks discovered the "golden mean," one of the most beautiful concepts of geometry.

The golden mean arose from a rather simple geometrical figure. The Greeks started with a rectangle of sides equal to 1 and 2. This uncomplicated rectangle can be cut in half by drawing a diagonal whose length is, using the Pythagorean theorem, $D^2 = a^2 + b^2$. The diagonal cut the rectangle in two parts, which in turn are right triangles.

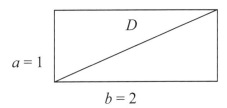

Substituting the numerical value for the sides of the rectangle, I find $D^2 = 1^2 + 2^2 = 5$, so the length of the diagonal is $D = \sqrt{5}$.

The Greek scholars disconnected one of the triangles on one corner of the $1 \times 2$ rectangle, straightened out the side equal to 1, and then rotated the side equal to 2 to form two sides of a new rectangle $2(\sqrt{5} + 1)$. By doing so, they discovered that the proportion or ratio between the two sides of the new rectangle is equal to a number with an infinite, non repeating decimal. They called this ratio $\phi$ (phi), defined as follows:

$$\phi = \frac{\sqrt{5} + 1}{2} \approx 1.6180339...$$

Fascinating! Ancient mathematicians also found that the ratio $\phi$ appears in many shapes in nature. Just like $\pi$ and $e$, the number $\phi$ seems to arise out of the basic structure of the universe. But unlike abstract numbers, $\phi$ appears regularly in the realm of living organisms that grow and develop in steps. For example, the Nautilus shell grows larger on each spiral by an amount $\phi$. Also, the sunflower has 55 clockwise spirals overlaid on either 34 or 89 counterclockwise spirals in a proportion equal to $\phi$. The number $\phi$ turns out to be an ideal rate of growth.

The Greeks were so impressed by this geometrical ratio that they applied it to many of their architectural designs and called it the divine proportion. The ratio $\phi = \dfrac{\sqrt{5}+1}{2} \approx 1.6180339...$ is now called the *golden mean.*

How amazing that a simple mathematical relation such as $\dfrac{\sqrt{5}+1}{2}$ can lead to a number that has no exact value, just like $\pi$. Does it mean that $\phi$ is also an irrational number? Yes, it meets the requirements, but to be sure I should prove it mathematically.

Are $\pi$ and $\phi$ related somehow? If so, how? The two numbers originate from a basic geometrical shape, both are a ratio of two numbers, and both are equal to a number that has infinite, non repeating decimals. Yes, $\pi$ and $\phi$ must be related. I like to think that the equation connecting these two special numbers must be beautiful and elegant.

That's one of the fascinations of mathematics. One can discover a myriad of relations of immeasurable beauty and worth. How can anybody not find mathematics marvelous?

Saturday—February 5, 1791

Geometry deals with many interesting problems that date back to the mathematicians of antiquity. The problems were challenging for the ancient scholars, not because they did not have solutions, or the solutions they had were unusually hard, but because the known solutions violated an important condition imposed by the Greek mathematicians themselves. For example, the problem known as the "doubling of the cube" requires the construction of a cube whose volume is double that of a given one.

The doubling of the cube is often referred to as the *Delian* problem because it deals with a legend related to the mythical oracle

120

at Delos. According to a letter from the mathematician Eratosthenes to King Ptolemy of Egypt, Euripides mentioned the *Delian* problem in one of his tragedies. The story relates that, in 430 B.C., the Athenians consulted the oracle at Delos hoping to stop the plague ravaging their country. Apollo advised them to double the size of his altar, which was designed like a cube. The Athenians tried but could not double the size of the cubic-shaped altar. As a result of several failed attempts to satisfy the god, the pestilence only worsened. Not knowing what to do, the Athenians turned to Plato for advice.

The Athenians could not solve this problem using only a straight edge and a compass, which were the only tools available to solve geometric problems like this one. The *Delian* problem challenged mathematicians for centuries, due to the restriction of using only those tools. Of course, the solution can be pursued with the help of algebraic analysis.

*Eh bien.* The doubling of the cube simply means that, starting with the volume of a cube, one must construct a cube whose volume is twice as great. Algebraically, this is reduced to finding a constructible solution to the equation $x^3 = 2a^3$, where $a$ is the length of the cube.

Can I simply show that the equation has no rational solutions? Assuming the opposite is true, let $x = p/q$ be a rational solution. Then $p^3 = 2q^3$. The number of prime factors on the left side of this equation is divisible by 3. The number of prime factors on the right side of the equation, when divided by 3, leaves a remainder of 1. Therefore, the equality is impossible indeed. This means that the equation $x^3 = 2a^3$ has no rational solutions.

Without this simple algebraic analysis, the Athenians could not double the cube!

Tuesday—February 15, 1791

I can't stop thinking about the *Delian* problem. The story asserts that the Athenians were advised by Apollo to double the size of his cubic-shaped altar. And they could not do it using rule and

121

compass. So, I restate the problem of squaring the cube (the altar) in the form of a binomial $x^3 - 2 = 0$, and try to find a rational solution.

Writing $x^3 = 2a^3$ implies that I have to construct the length $x$ for the cube given by $x = a\sqrt[3]{2}$. If I can find a rational solution, where $x$ and $a$ are integers with no factor in common, I should find that $x^3 = 2a^3$ is an even integer. Since the cube of an even integer is even, $x$ must also be even, ($x = 2p$), which gives $x^3 = 8p^3 = 2a^3$, and $a^3 = 4p^3$. That is, $a^3$ is an even integer and, consequently, $a$ is also even, and $a$ and $x$ have 2 as a common factor, which contradicts my original hypothesis. Therefore, I can only conclude that the binomial $x^3 - 2 = 0$ is irreducible over the field of rational numbers. In other words, $a\sqrt[3]{2}$ is an irrational number. Just as I had deduced before! Did the Spanish girl arrive at the same solution?

Oh, *mon Dieu*! It's so late. I better get under the covers or Maman will be so angry!

Monday—February 28, 1791

France is in turmoil. Some people oppose King Louis XVI; others support and defend him. A violent confrontation of the two sides erupted today, revolutionaries against royalists, right in front of the Tuileries Palace. It was scary. The Marquis de La Fayette, the commander of the National Guard, was away attempting to disperse riotous mobs in the countryside.

At the same time, a group of several hundred aristocrats and supporters of the court, concerned for his safety and armed with guns and knives, gathered and stood guard around the palace. Suspecting that this demonstration was an attempt to help the king escape, the revolutionaries started a violent confrontation. La Fayette had to return in a hurry to disarm the crowd and arrest many of them. Papa was not involved, as he was busy at the meeting of the Assembly. But when we heard the commotion in the streets, we were worried for Papa's safety.

Paris is in a state of chaos. But at home we try to preserve a sense of normality. We maintain the same family routines, including the literary evenings and musical interludes that I enjoy so much. I occupy the rest of the time with my studies. And now I have a new activity that brings me much pleasure. I am teaching Millie to read and write. Last night she copied a few paragraphs from a book I gave her for her birthday. She is an eager pupil, so I taught her basic arithmetic. Millie can count and do simple adding and subtracting, but I will teach her to calculate percentages.

Thursday—March 10, 1791

The pope threatened excommunication to all the priests who swear the civil oath. Maman thinks that the decree of His Holiness may stop the madness sweeping France. Swearing an oath to a civil constitution renegades on the most sacred principles. I don't think the message from the pope brings comfort, on the contrary. His threat puts the priests in a more difficult position, as they have no choice. If they do not abide by the new constitution, the priests will be arrested, imprisoned, and perhaps sentenced to death. Mother is worried but adamant; she will not take communion from any of the government priests because she feels they have betrayed the Church. What a predicament for the people like my mother and for the clergy.

I have my mathematics to fulfill me. There is so much to learn! Mathematical analysis requires deep concentration. It also requires that the mathematician observe nature with the eyes of the intellect in order to see how it works. The Greeks knew this. They discovered, for example, that any rectangle constructed to the proportion of the Golden Mean holds the *golden spiral* coiled within it. Unlike a regular spiral, the distance between the coils of the golden spiral keeps increasing, growing wider as it moves away from the center. Mathematicians see this as the spiral of constant expansion and growth. And it is found in nature, in many living creatures, from seashells to sunflowers. As one mathematician wrote, "The golden spiral is the template of growth, the mathematical formula for evolution." And as I say, mathematics is the most wonderful and poetic way to discover the universe!

Angelique draws beautifully. I told her she should be an artist. With her charcoal sketches she captured the beauty of the palaces and buildings in the neighborhood. Now she wants to draw people, and she did a very good job with her first attempt. Millie volunteered to pose for a portrait. I must admit it; the pencil sketch depicts the gentleness of Millie's eyes and her sweet smile. Angelique just had a little problem drawing her nose. Now she is after me, but not meaning to dampen her enthusiasm, I suggested she draw our mother's beautiful face instead and give her the sketch as a birthday present. My sister thought it was a great idea and left me alone.

I am now learning theorems and proofs. What do mathematicians mean when they talk about theorems? Different books say different things, but in simple terms, a theorem is a statement or formula that can be deduced from the axioms of a formal system, by recursive application of its rules of inference. An axiom is a statement that is stipulated to be true for the purpose of constructing a theory in which theorems may be derived by its rules of inference. Well, these are too many words. The best way to illustrate a theorem and its proof is by doing it. First of all, to prove a theorem I need to know certain facts, and then I must follow appropriate mathematical rules.

I start with a very easy mathematical statement: "The sum of any positive number and its reciprocal is greater or equal to 2." This is a theorem, written in my own simple words. Thus, I must restate it using mathematical notation:

$$\text{If } x > 0, \text{ then } x + \frac{1}{x} \geq 2$$

This is how I can prove that this statement is true:
Since $$(x - 1)^2 \geq 0$$

Therefore $\qquad\qquad x^2 - 2x + 1 \geq 0$

or $\qquad\qquad\qquad x^2 + 1 \geq 2x$

Hence, dividing by $x$, I get $\quad x + \dfrac{1}{x} \geq 2$

Q.E.D.

*Eh bien*! This theorem is easy to prove because I knew that the square of any real number is nonnegative, and that adding equal quantities to both sides of an equation preserves the inequality.

That's what makes a mathematician. To prove theorems one must know enough mathematics. Also, when mathematicians finish the proof of a theorem they write Q.E.D. as the last statement. When I began to study the book *Introduction to Analysis*, I wondered why Euler always finished his proofs with Q.E.D. Now I know that the three letters are an abbreviation of an expression in Latin, *Quod erat demostrandum*; it means, "Which was to be demonstrated." From now on, when I finish a proof I will also write Q.E.D.

Friday—March 25, 1791

Mathematics can be challenging. Sometimes mathematicians sense that something is true, or predict an outcome, but they cannot prove it. A learned person can see beyond what is before them and get a notion of what is or could be there. Take for example Christian Goldbach, a mathematician who studied prime numbers. He saw mathematical relations that were not obvious to others. For example, he surmised that every even integer greater than 2 can be represented as the sum of two primes. Goldbach also conjectured that every odd integer is the sum of three primes.

Goldbach did important work in number theory, much of it in correspondence with Euler. He traveled throughout Europe meeting the greatest mathematicians of the time, such as Gottfried Leibniz, Nicolaus Bernoulli, de Moivre, and Daniel Bernoulli.

Goldbach also conjectured that every odd number is the sum of three primes. His statement is called a conjecture because he predicted something without having all the supporting evidence. This type of work is important to mathematicians because it challenges them to prove or disprove the statement he made.

I, too, would like to make conjectures and then prove them without doubt. I would be happy to simply prove a conjecture or theorem that nobody else has been able to prove. Am I intelligent enough to do it? I lack the knowledge to do it just now. I must learn much more.

Sunday—April 3, 1791

The town crier woke us this morning with unexpected news, "Mirabeau is dead!" We ran to the windows to hear the proclamation. The court official went from corner to corner, yelling the distressing news through the neighborhood. Mirabeau, the president of the National Assembly, died last night.

The Comte de Mirabeau was a friend of my father's and came to our home a number of times to discuss politics. Monsieur Mirabeau was one of the supporters of a constitutional monarchy, and he had tried to reconcile the reactionary court of Louis XVI with the increasingly radical forces of the revolution. Papa says that Monsieur Mirabeau was in bad health and was under a lot of stress since his election. Although my mother did not particularly like his questionable lifestyle, she is sorry to learn of M. Mirabeau's passing. There is no one else in the government as prominent and influential as Mirabeau was to speak on behalf of King Louis.

Thursday—April 7, 1791

I wish to find a method to predict primes without having to perform endless arithmetic calculations. But it is not easy. Ever since the discovery of prime numbers, many mathematicians have attempted to formulate a method to generate primes.

In 1640, Pierre de Fermat developed a formula that he thought would produce prime numbers. He stated that numbers of the form $2^n + 1$, where $n = 2^p$ and $p$ is a positive integer, would be prime numbers. However, he was wrong.

It is easy to show that Fermat's formula works for integers 0, 1, 2, 3, and 4, but it fails for $n = 5, 6$, and other values of $n$. For $n = 32$ (or $p = 5$), the formula gives (according to Euler) $2^{32} + 1 = 4294967297$, a number that can be divided by 641, therefore it is not prime. I do not have the energy or time to calculate the power $2^{32}$, so I cannot verify Euler's value. Of course I believe it, but I wonder how he did it, considering that he was partially blind. It is true; he had a phenomenal ability to do complex calculations in his head. Still, this number is too huge to calculate it this way.

A few years after Fermat, a colleague of his named Marin Mersenne published a formula for prime numbers of the form $2^n - 1$, where $n$ is a prime number. I do not know if Mersenne implied that all numbers of this form would be primes, so I checked it for three values of $n$ arbitrarily chosen: $n = 2, 3$, and 11. Well, the first two are indeed prime numbers, but not the last one since, for $n = 2$ and 3, the formula gives $2^2 - 1 = 3$ and $2^3 - 1 = 7$. However, for $n = 11$ it yields $2^{11} - 1 = 4095$. The number 4095 is at least divisible by 3, so it is not a prime number. This tells me that Mersenne's formula does not generate prime numbers for all values of $n$.

Has anyone else developed a different formula since then?


Sunday—April 24, 1791

We went to church early in the morning to celebrate Easter, and in the afternoon we visited M. and Mme Geoffroy. We had dinner with them and a lovely musical soirée. My sisters sang the solemn *Te Deum* to the delight of my parents. A friend of Papa came later with disturbing news. He told us what happened to the royal family this morning.

This morning, King Louis and his family were trying to go to his château in Saint Cloud. But the National Guard, flanked by a crowd of angry citizens, stopped the royal carriage at the palace gate. The king told the soldiers they wanted to celebrate Easter Mass in St. Cloud's chapel, but the soldiers did not allow the carriage to leave. The family wanted to take Easter communion from their own loyal priest. Their reluctance to take the sacrament from a juror priest makes the king suspect in the eyes of the government. It is not a good sign.

Saturday—April 30, 1791

Madame de Maillard is one of the most sophisticated and educated women I know. Her knowledge of literature and philosophy is vast. When she speaks, she does it with such passion that one cannot help but listen to her point of view. I admire her confidence and her eloquence.

Mme de Maillard believes that women should have the same rights as men, politically and otherwise. She argues that if women had political power, we could change the social system in such a way that it would benefit everybody, not only other women and children. Mme de Maillard is the only woman who doesn't think it wrong for me to study mathematics. In fact, she encourages me. She gives me books and tells me stories about ancient philosophers. Mother no longer chastises me for studying mathematics; I believe her tolerance is a result of listening to Mme de Maillard, who gives very convincing arguments as to why women should be educated. She reasons that society would benefit greatly if as many women as men would go to the universities, but she admits there is a lot to be done to change society's notions about the role of women.

I personally do not care for political power. However, without civil rights, women do not have the power to seek and receive the same type of education as men.

The Civil Constitution of the Clergy now puts the Church under the control of the State. At tonight's meeting, my father and his friends talked about the new developments. Many priests were arrested because they refused to swear the oath. I just sat silently in my corner with a book before me, although unable to concentrate, thinking about the foolishness of the government's actions. The revolution started with the idea of social equality, so what does that have to do with priests? How is an oath going to uphold those principles? I can understand the appropriation of lands and church treasures, and fighting the corruption among the powerful church leaders. But imprisoning priests for not accepting to be treated like workers of the government? That is not fair.

After their political discussion, Papa and his colleagues talked about France's scientific position in Europe in the last two centuries. They mentioned the great philosophers and scholars Pascal, Descartes, Mydorge, Mersenne, Desargues, L'Hôpital, and Fermat. One of Papa's friends mentioned the work of contemporary French mathematicians, some of which are also involved in the political affairs of the new government.

They mentioned Lagrange, Laplace, Condorcet, Monge, Legendre, and Carnot. Papa knows the marquis de Condorcet very well. In fact, he is the one who sent me the two volumes of Euler's *Introduction to Analysis* last year. M. Condorcet is also known for his activism concerning women's rights after he wrote the essay *On the Admission of Women to the Rights of Citizenship*. He believes that women must share identical educational rights with men. If I remember correctly, many of Papa's friends were against M. Condorcet's ideas regarding women, but in my opinion he is better than most. A man must be quite intelligent in order to accept that a woman is his intellectual equal.

I would like very much to learn the mathematics of the scholars mentioned at the meeting. One day I will muster the courage to write a letter and request permission to attend their public lectures at the Louvre.

*O*nce upon a time, a great man named Don'Raf wandered through a distant land accompanied by his young son. One afternoon, they came upon a stranger sitting by the side of a deserted road. The man was lost, stranded without food, and so Don'Raf offered to help him find his way home. The foreigner accepted and asked for something to eat. Don'Raf responded that he had five loaves of bread and his son had three. Combined, the eight loaves would be shared among the three of them. Thanking them, the stranger promised to recompense them for their generosity with eight golden coins as soon as they reached his home. After eating, the three men mounted the horses and began their journey back to the city.

The next day, upon arrival, Don'Raf and his son were surprised to learn the stranger was the king of that territory. The grateful monarch invited his benefactors to stay the night in his palace. After a splendid supper, he ordered his servant to give the boy three gold coins for the three loaves of bread he shared with him, and five to the father.

But Don'Raf interjected. "Although your division of the coins seems straightforward, it is not mathematically correct. Since I shared five loaves, I should be paid seven coins; my son, who shared three loaves, should get one coin." The king was bewildered, for he did not understand the logic of such argument. Don'Raf offered to demonstrate why his proposal was mathematically correct:

"During the trip, when we were hungry, I took one loaf of bread and divided it into three parts, and each one of us ate a piece. If I contributed five loaves, that means there were fifteen pieces of bread from my bag, right? And my son contributed three loaves, or nine pieces from his. There were a total of twenty-four pieces of bread, so each one of us ate eight pieces. Now, of the fifteen pieces from my bag, I ate eight, which means I gave away seven pieces. My son had nine pieces and ate eight, so he only gave away one piece of bread. That is why I expect to receive seven coins and my son one."

The king could not argue against such logic; therefore, his attendant distributed the eight coins between father and son *according to the father's mathematical formula. Don'Raf thanked the monarch and added: "Your majesty, the division that I proposed, seven coins for me and one for my son, is mathematically correct, but it is not fair in the eyes of God." And gathering the eight coins, the wise man divided them into equal parts: four coins for his son, and four coins he kept.*

I composed this tale to entertain my little sister, and also to remind myself that life does not follow an exact mathematical formula.


Saturday—May 28, 1791


Scholar Marin Mersenne contributed to the development of prime numbers. I've searched in several books, trying to learn about the person behind all that analysis, and this is all I was able to gather: Marin Mersenne was a theologian who corresponded with other eminent mathematicians about his studies, and he played a major role in communicating mathematical knowledge throughout Europe at a time when there were no scientific journals.

Mersenne received many visitors in the monastery here in Paris, most notably scholars like Fermat and Pascal. Mersenne defended Descartes and Galileo against theological criticism, and struggled to expose the pseudo sciences of alchemy and astrology. He expanded Galileo's work in acoustics, and with his interest in music, Mersenne provided inspiration and mentoring to the Dutch scholar Christiaan Huygens, the one who wrote *Theory of Music*. Huygens planned to move to Paris in 1646 to be near Mersenne and expand their collaboration. However, the two never met, as Huygens arrived several years after Mersenne had died. In number theory, Mersenne's work is very important. He tried to find a formula that would represent all primes but, although he failed in this, his work on numbers of the form $2^p - 1$, where $p$ is a prime number, stimulated further developments in number theory.

*Oh Dieu*, it is late and my eyes feel tired. I'll continue this line of thought tomorrow.

131

I've solved few problems lately. I feel a little lethargic and physically tired; perhaps it's due to the weather. It is not summer yet but the air feels hot and humid. I spent the day reading biographies. It is fascinating to learn how mathematicians lived, how they studied and developed their mathematics. I started with the *Dictionnaire raisonné des sciences, des arts et des métiers*. My father showed me this encyclopedia when I was a little girl, but I do not remember reading about mathematics at that time. I also found a book about the great scholars of France in the last two centuries.

There was a mathematician who was a child prodigy in mathematics. His name was Guillaume François Antoine Marquis de L'Hôpital; he was born in Paris in 1661 and died in 1704. When he was fifteen years old, he heard two mathematicians discussing one of Pascal's problems. L'Hôpital surprised them when he stated that he could solve the problem. Indeed, a few days later, he came back with the solution. As the son of the lieutenant general of the king's armies, he was intended for a military career, but L'Hôpital was forced to resign because he had a vision problem. He then devoted himself entirely to mathematics, his favorite subject. When he was thirty years old, L'Hôpital learned calculus from Johann Bernoulli, another great mathematician.

L'Hôpital was a very good mathematician and became better known after he wrote in 1696 a book titled *Analyse des infiniment petits pour l'intelligence des lignes courbes* (*Analysis of the infinitely small to understand curves*), which was the first textbook on differential calculus. In the introduction the Marquis L'Hôpital acknowledges his mentors Gottfried Wilhelm von Leibniz, Jacob Bernoulli, and Johann Bernoulli who taught him the calculus. L'Hôpital had the greatest teachers!

Saturday—June 18, 1791

The word *brachistochrone* is associated with a famous mathematical problem posed by Johann Bernoulli in 1696.

Mathematicians use the brachistochrone to describe the curve between two points, which, for a given type of motion, represents the trajectory that takes the shortest time. An example of this motion is given by the fall of bodies due to the force of gravity. Prior to Bernoulli, Galileo claimed that the natural descent of a body takes the shortest time along the arc, which connects the ends of a chord joining two points on a circle. But apparently he was wrong. Bernoulli reintroduced the brachistochrone problem in the scientific journal *Acta eruditorum*.

Bernoulli stated the problem as follows: "Given two points A and B in a vertical plane, what is the curve traced out by a point acted on only by gravity, which starts at A and reaches B in the shortest time?"

Johann Bernoulli knew the solution of the brachistochrone problem, but he challenged others to solve it, too. Mathematicians thrive on challenging not only themselves, but others as well. I've read about this kind of contest throughout the history of mathematics. The challenges are placed to test the mathematicians' methods and the strength of their intellect.

Gottfried Leibniz persuaded Johann Bernoulli to allow a longer time than the six months he had originally intended for solving the brachistochrone problem. Leibniz reasoned that allowing more time would give foreign mathematicians a chance to solve the problem. Several solutions to the brachistochrone problem were submitted by the best mathematicians of the time: Isaac Newton, Jacob Bernoulli, and Leibniz, in addition to Johann Bernoulli himself. They all proved, independently, that the path of least time is an inverted cycloid.

I am curious to see how they did it. Originally I guessed, like Galileo, that the curve traced by a point acted on by gravity would be an arc of a circle to move in the shortest time. Now I know that it is not. I must see what type of equation is used to solve problems like these. I will ask Papa to take me to the city library to consult the *Acta eruditorum* and review the solutions that these mathematicians submitted.

The king attempted to escape and was caught. This caused a big commotion in Paris. And I witnessed how the royal family was escorted back after their capture. Yesterday, while riding with Mother and Angelique, our carriage had to stop at the Pont Neuf, as a large group of people blocked the bridge. We did not know what the uproar was about until later in the evening, when Papa came to dinner and told us.

A few days ago Louis XVI, Marie Antoinette, and their children attempted to flee France. The royal family was arrested at Varennes and brought back to Paris. That's what was happening when our carriage was blocked. As we were waiting by the bridge, the royal coach passed us, escorted back to the Tuileries palace. The mob was hysterical! Many people taunted the royal family; men beat on the carriage windows with sticks and pikes.

As the vehicle went by, the men kept their hats on in a gesture of disdain that is so disrespectful. People shouted insults, using such foul language that it made my mother cringe. She kept muttering with tears in her eyes, "Messieurs, in the name of heaven! He is the king of France, for God's sake! Please be kind to her; she is the queen!" As we got closer, we witnessed the rude soldiers pointing their muskets to the ground as the royal family was taken back to the palace. Mother was appalled and deeply troubled by the way people behaved towards the monarchs.

Papa says that after the king attempted to flee France, many people, even those who love and support the monarchy, now think that Louis XVI is a traitor and a danger to the nation. The discussion with his colleagues centered on the suspension of the king's authority by the Constituent National Assembly. There is now increased suspicion of royal conspiracies and foreign invasions.

Mother supports King Louis and sympathizes with his attempt to leave a country that no longer respects him. She is troubled to see that the reverence for His Majesty was replaced by contempt, and that the traditional standards that bounded the French people to the king are destroyed. However, one friend of my father clarified that a group of people pasted posters all over, ordering the citizens to keep their hats on when the royal carriage passed by, threatening them with arrest if they didn't. So, it is difficult to say whether the citizens lost their respect for the king, or the leaders of the revolutionary movement are simply manipulating them. Either way the monarchy is crumbling.

Tuesday—July 5, 1791

I attempted to solve some problems, but try as I might, I could not find the answers. I was about to give up when, quite by accident, I stumbled upon a book whose author claimed that one cannot be a real mathematician if one does not master Euclid's *Elements*. I assumed that this ancient book delved mainly on geometry, a part of mathematics I thought rather boring because it requires the measurement of geometrical shapes with a ruler and compass. Still, in an effort to overcome the obstacles that prevented me from solving these equations, I decided to spend some time studying Euclid's geometry. Papa bought me a copy of Euclid's *Elements*.

I found the *Elements* to be a wonderful book. Its beauty lies in its logical development of algebra and geometry. The *Elements* is a set of thirteen books. The first book contains definitions that are fundamental to the study of the mathematics that follows. The first few terms in the *Elements* are primitive terms. Their meaning arises from properties about them assumed known later in the axioms, which are of two kinds: postulates and common notions. Euclid defines a point as "that which has no part." This means that a point has no width, length, or breadth, but has an indivisible location. The first postulate, I.Post.1, for instance, gives more meaning to the term "point." It states that a straight line may be drawn between any two points.

"Line" is the second primitive term in the *Elements*, defined as a set of consecutive points with breadthless length. The description "breadthless length" implies that a line will have one dimension, length, but it does not have breadth or depth. I cannot tell from this definition what kind of line is meant by "line," but later a "straight" line is defined to be a special kind of line. I conclude, then, that "lines" need not be straight. Perhaps "curve" would be a better translation than "line," since for Euclid a curve may or may not be straight. Later, Euclid defines new concepts in terms of others defined earlier.

The subject matter of Book II is called geometric algebra. Since the geometry is plane geometry, the relations represent quadratic equations. This should be easier for me to understand since I know the basics of algebra.

Monday—July 11, 1791

How can a man who died thirteen years ago be so popular? For the past several weeks people talk about Voltaire as if he were alive. Voltaire was a writer and philosopher, an outspoken critic of the monarchy and of the prelates in the Church. Voltaire's ideas and thinking are now very popular. Many theaters are staging his plays, and his name is invoked whenever someone wishes to justify his opposition to the monarchy.

Because of his criticism of the Church, Voltaire was denied burial in church ground in Paris when he died in 1778. Now, at the urging of the Jacobins—the political leaders who oppose the king—Voltaire's remains were transferred to the *Panthéon*, a monument to the *Grands Hommes,* the great men of France.

The coffin was placed at the ruins of the Bastille since last night. And today we watched the funeral procession on Rue Saint Honoré. It was led by a cavalry troop, followed by delegations from schools, clubs, fraternal societies, and groups of actors from the theaters. The workers who demolished the Bastille carried balls and chains from the prison. Four men dressed in classic theater costumes carried a golden statue of Voltaire. Actors waved banners inscribed with the names of Voltaire's plays. A full orchestra preceded the wheeled sarcophagus, which was drawn by twelve white horses. The casket was decorated with theater masks, and the inscription, "Poet, philosopher, historian, he made a great step forward in the human spirit. He prepared us to become free."

Many important people trudged solemnly behind the sarcophagus, including members of the National Assembly, the judiciary and the municipality of Paris. Hundreds of citizens lined the streets to watch the funeral procession, and many people followed it. We did not go to the cemetery, but Papa said that it was almost midnight when Voltaire's remains were placed to rest at the *Panthéon*.

I am familiar with Voltaire's philosophy, a topic of discussion at home since I was a child. Maman did not like his ideas because Voltaire criticized the king, but Papa agreed with his writing against religious intolerance and persecution. In fact, sometimes Papa reads Voltaire's plays during our family literary evenings. One of his favorites is *Oedipe*. Papa also enjoys his humorous poetry, an art that Voltaire used to poke fun at the government and nobility.

At one time, Voltaire was imprisoned in the Bastille for writing a mocking satire of the government. After he insulted the powerful nobleman, Chevalier De Rohan, Voltaire was given two options: imprisonment or exile. He chose exile and went to live in England. When he came back to Paris, Voltaire became friends with the Marquise du Châtelet, a very intelligent lady. They studied natural sciences and wrote books together.

Voltaire continued writing books and plays after the death of Madame du Châtelet. Voltaire introduced notions of equality and tolerance between men that went beyond social and religious prejudice. Thus, although he is dead, Voltaire's name and his works will live forever.

Friday—July 15, 1791

Mathematicians create mathematical ideas and fascinating worlds independent of the everyday world that we inhabit. Mathematical ideas are figments of the mathematician's imagination, produced by sheer logic and splendid creativity. From Euclid I learned that a point only shows location, but it cannot be seen since it has zero dimensions. Yet, we can see a line segment composed of these invisible points. Euclid also defined a line as infinite in length, but does such a figure exist in the realm of our lives? What about a geometric plane, defined as infinite in two dimensions and only one point thick? These mathematical objects can only exist in another world, the world of mathematics.

However, many mathematical ideas are meant to connect to our world, even explain it. An equation can represent something real, like the motion of a body or the change in temperature; that is why I think of mathematics as magic. Is it enough to know mathematics as a pure intellectual recreation? I think not. Knowing to solve mathematical puzzles is very entertaining, but I want to uncover the secrets of the universe. That is why I must learn other sciences.

If I wish to understand nature, I must know the laws that govern it. Science teaches me the laws; it incorporates basic ideas and theories about how the universe behaves. Science also provides me intellectual pleasure, giving me a framework for learning and pursuing new questions, and gives me an unparalleled view of the magnificent order and symmetry of the universe. Mathematics is the language of science, thus I need to know both.


Wednesday—July 20, 1791


The second anniversary of the destruction of the Bastille was last week. And, tragically, it was marked by a violent confrontation that ended in a horrible massacre, again! The fanfare of last year was missing; there was no royal ceremony, no special outdoors Mass for all the citizens of Paris. This time the sans-culottes held a demonstration on the Champ de Mars to gain signatures for their petition to depose King Louis XVI. The demonstration became violent. A contingent of National Guards, the soldiers led by General de La Fayette, fired on the crowd, killing at least fifty demonstrators. This enraged the leaders of the revolutionary movement and fueled their hate for the monarchy.

Right now the power of the king is suspended until he signs the Constitution. The Assembly is divided. Many deputies support a democratic government with a ruling king. However, the citizens are divided in their support. Right after the Assembly voted to reinstate the powers to the crown, radical groups of sans-culottes began circulating petitions for his dethronement. Papa believes that, instead of promoting peaceful resolutions, the massacre at the Champ de Mars intensified the resolve of the sans-culottes to destroy the monarchy.

As Maman says, political debates can turn friends into enemies. I have seen grown men behave like quarrelsome children, shouting at each other when they disagree about some issue. I should be immune to these verbal matches. I've grown up listening to my father and his friends arguing when they meet every Tuesday night in his library. Lately the discussion is so heated that I no longer want to be around. Some of the men argue in favor of a constitutional monarchy. Others believe that France should become a republic without a king. My father tries to defend Louis XVI, but they all wonder why the king fled, and some claim that this act betrays the people and the nation. Even those who oppose a republic are divided on their support for the monarchy.


Friday—August 5, 1791


Paris is under martial law. There are soldiers everywhere, watching everybody, ready to use their firearms on any citizen attempting to revolt. The soldiers stop people in the streets for no apparent reason, and if suspected of conspiracy, a person can be taken to prison. I feel depressed and confused. There is no rational justification for the bloodshed and the horrors that ravage my country. I spend every day reading the works of great philosophers, trying to understand human behaviour, trying to make sense of the senseless. But nothing justifies the violence and rampant lawlessness.

Paris seems dark and sinister, a city where innocents and guilty alike fear for their lives. After the tragic events of last week I've became painfully aware of the disparity in the revolutionary ideology. France is divided into two groups. The moderates support a democracy without removing the king as the head of the nation. They believe in peace and wish to put an end to the hostility and brutality of the past two years. On the other side are the radical groups who demand the dethronement of King Louis. These radical revolutionaries are ready to use any means, including murder, to achieve their goals. Martial law keeps these citizens' anger from boiling over.

My parents try to keep a sense of normality at home, trying not to dwell too much on these awful times. We still have our family gatherings every week. That is the only way to keep us from fretting too much about the events of which we have no control. I am already preparing for tomorrow's get-together. I will talk about ancient Greek philosophers and mathematicians.

In particular, I will discuss Euclid and his contributions to science and mathematics. I wish I could also describe the man. But very little is known about Euclid's life. According to Proclus, Euclid came after the first pupils of Plato and lived during the reign of Ptolemy I. Thus, historians believe that Euclid was educated at Plato's academy in Athens, and that that he lived sometime around 300 B.C.

On the personal side, Euclid was probably a kind and patient man. I read an anecdote that reveals a little of his character. The story describes how one day, one of Euclid's pupils after his first geometry lesson asked what he would gain from learning geometry. Euclid told his assistant to give the pupil a coin so he would start gaining from his studies. Very clever, indeed! In another story, King Ptolemy asked Euclid if there was an easier way to learn geometry. At this Euclid replied, "There is no royal road to geometry," and sent the king to study!

When Euclid wrote *Elements*, he synthesized all the available knowledge on geometry of his time. Euclid also proved that "the number of primes is infinite" and gives a beautiful proof of this theorem. He used the methods of exhaustion and *reductio ad absurdum* to prove many theorems. He also discussed the algorithm for finding the greatest common divisor of two numbers.

I could not find anything about the circumstances of Euclid's death. All I know is that he lived and worked in Alexandria for much of his life, and that he established a mathematical school in Alexandria. It is not much of a biography, but I think this is enough for my presentation about Euclid, one of the greatest and most influential mathematicians in history.

Friday—August 12, 1791

The following is a theorem in the second book of the *Elements*:

*If a straight line is cut at random, the area of the square on the whole line is equal to the areas of the squares on the segments plus twice the area of the rectangle formed by the segments.*

I translate the statement into an algebraic equation:

$$(a + b)^2 = a^2 + 2ab + b^2$$

How did Euclid establish such relationship? His proof is very complicated, but I can prove it in a simpler way, if I invoke some rules of algebra:

$$(a + b)^2 = (a + b)(a + b) = (a + b)a + (a + b)b$$
$$(a + b)^2 = a \cdot a + b \cdot a + a \cdot b + b \cdot b$$
$$(a + b)^2 = a^2 + 2ab + b^2$$

*Très bien.* Now, how do I generalize the equation to higher exponents, to obtain a formula for the expansion of $(a + b)^n$ for $n = 3, 4, ..., n$?

The first step should be to obtain an expansion for $(a + b)^3$. I could not find a theorem similar to II.4, giving the expansion of $(a + b)^3$ in any of the Euclid's books. So, following the same method as I did before I get:

$$(a + b)^3 = (a + b)(a + b)^2 = (a + b)(aa + ab + ba + bb)$$
$$= a\,(aa + ab + ba + bb) + b\,(aa + ab + ba + bb)$$
$$(a + b)^3 = aaa + aab + aba + abb + baa + bab + bba + bbb.$$

141

or

$$(a + b)^3 = a^3 + 3a^2b + 3ab^2 + b^3$$

This is a natural progression to that one can follow to arrive at the more general $(a + b)^n$. There must be an easier way to develop the expansion of $(a + b)^n$ rather than the algebraic method I used for $n = 2, 3$. But how?

Friday—August 19, 1791

I've wondered how to prove that there are an infinite number of primes. Euclid thought about that too and proved it. In his *Elements*, Euclid gives a proof to show how. He did it by first assuming that there are finite numbers of primes, and then proving that statement to be false. The proof is as follows:

Assume there are a finite number of primes $p_1, p_2, p_3, \ldots, p_n$, then $p_1 p_2 p_3 \cdot \ldots \cdot p_n + 1$ is not divisible by any of the $p$'s, so any of its prime divisors yields a new prime number.

Euclid only considered the case $n = 3$. Just like Euclid, I can show that $p_1 \cdot p_2 \cdot p_3 + 1$ is not divisible by any of the $p$'s.

Taking the first three primes:  $2 \cdot 3 \cdot 5 + 1 = 31$

Indeed, 31 is not divisible by 2, 3, or 5. However, 31 is a prime number.

But I can't do this for all numbers. I will try the proof differently.

Suppose $n$ represents the last prime number. Then I form a new number from the product of all prime numbers up to, and including, the last number $n$: $2 \cdot 3 \cdot 5 \cdot 7 \cdot 11 \cdot \ldots \cdot n$.

Now I add 1 to this product and call it $k$, that is, $k = 2 \cdot 3 \cdot 5 \cdot 7 \cdot 11 \cdot \ldots \cdot n + 1$. $k$ must be prime. If $k$ were not prime, then the list of primes I used to form the product must be missing one. I know that 2, 3, 5, 7, 11, ... $n$ cannot divide evenly into $k$ because

142

there would always be one left over every time I tried to divide by any of the numbers 2, 3, 5, 7, 11, ..., $n$. Therefore, $k$ must be a new prime. This means that the primes are infinite indeed!

Euclid is correct. I could prove his theorem with any number of primes using the same approach. However, that would be too tedious and unnecessary. I'll search for another more rigorous approach to prove that there are infinite primes.

Tuesday—August 30, 1791

I discovered something amazing. I discovered that between the integers 1 and 2 there are at least three irrational numbers: $\sqrt{2}$, $\phi$, and $\sqrt{3}$. Between 2 and 3 there are at least five irrational numbers: $\sqrt{5}$, $\sqrt{6}$, $\sqrt{7}$, $\sqrt{8}$, and Euler number $e$. And between the integers 3 and 4 there are at least seven irrational numbers: $\pi$, $\sqrt{10}$, $\sqrt{11}$, $\sqrt{12}$, $\sqrt{13}$, $\sqrt{14}$, and $\sqrt{15}$.

The three numbers $\phi$, $e$, and $\pi$, are the most extraordinary irrational numbers I know, each located between two consecutive integers, and the three are found between numbers 1 and 4. Between 4 and 5 one finds the square roots of 16, 17, 18,... and 24. But I wonder if there is another *unique* and *special* irrational located between 4 and 5? If so, it must be a number that relates to something beautiful in the universe. Oh, I am sure of it. Some brilliant mathematician will discover it one day. Or maybe I will find it myself! If I do, I will call it $\alpha$ or $\omega$.

This evening Monsieur de Maillard asked why I study mathematics. I was taken aback by the question and I blurted out, "Mathematics is the most beautiful science!" He looked at me strangely, smiled, and left to join my father. I wanted to explain more, to say that I like mathematics because I enjoy solving equations, and because I take pleasure in pondering the mysteries that lie beneath each theorem. I am intrigued and mystified by numbers and their harmonious connections in equations, like the notes in a beautiful symphony. Yes, mathematics is a rigorous science that demands logic and truth, because it is meant to solve the problems of the universe, including the everyday world around us and the one beyond our reach. I like mathematics because mathematics is truth!

At least I wanted to say that mathematics is a language that speaks to me in beautiful tones. However, I am too shy to express these feelings and thoughts to anyone. I only know that when I study mathematics, I transport myself to another world, a world of exquisite beauty and truth. And in that world I am the person I like to be.

Thursday—September 8, 1791

I woke up at the sound of thunder. Rain is coming, and with it a relief to the hot summer days that refuse to end. The lightning made me nervous so I decided to write to calm me down.

Last night I saw Antoine-August again. Maman gave a dinner party and invited M. and Mme LeBlanc, who showed up with him. I had not seen Antoine in a long time and wanted to know how he was, but I was too shy to ask. Papa, however, inquired about his studies, as if reading my mind. Mme LeBlanc replied that Antoine-August lost interest in mathematics and would rather study law. Well, that's discouraging.

I know there is much to learn, and there is no one to tell me what topics I must study. However, there is still some hope. Monsieur LeBlanc whispered to my ear that he will insist that his son pursue an engineering degree. Papa and M. LeBlanc commented that Antoine must master mathematics and science in order to pass the rigorous entrance examination. Only the best students are admitted to the *École des Ponts et Chaussées*.

I wish I could be the one preparing for the entrance exams. I would study very hard trying to be the best student in the class. But I know women are not allowed in engineering schools. What a pity. Not that I wish to become an engineer, but I would like to study there or any other school to attend the lectures in science and mathematics given by the best professors in France. For now going to school and being instructed by great mathematicians remain only a dream, a fancy of mine that sustains me through the long days, waiting for a better future.

Wednesday—September 14, 1791

King Louis XVI accepted the new Constitution of France. The National Assembly's document provides for an executive—the king—as well as a legislative body. The new Constitution is a moderate document that creates a constitutional monarch and gives the wealthy citizens privileges to a considerable degree.

Papa, who was present at the signing, described a ceremony that took place without the pomp and fanfare that would be expected of such historical event. He said that King Louis sat in an armchair, not a throne, symbolically placed at the same level as that of the president of the Assembly. Papa remarked that the deputies of the National Assembly kept their hats on during the ceremony, and that Louis XVI was addressed as "sir", instead of "His Majesty."

King Louis, who was raised to believe himself semi-divine, was treated like any other man. Louis XVI is no longer the beloved monarch of yesterday. As Papa says, the rapport between the king and the people was broken. Nor even the military respect the queen. The other day, I spotted a group of soldiers strolling in the Tuileries gardens with their hats on in the presence of Her Majesty. Some soldiers even sing disgusting songs as they approach the gates of the palace, as if aiming to humiliate Marie Antoinette. It is not only insolent, but also hurtful, especially because the soldiers use such vulgar language in front of the children. I am so embarrassed and wish I never hear that.

Now that the king signed the new Constitution of France, the country may regain some stability. At least today all bitterness was replaced by celebration. There was a spectacular pyrotechnic display this evening, which I saw from my window. I enjoyed the multicolored explosions that lighted the night sky. The citizens of Paris are laughing again, dancing in the streets full of hope for a better future.

Is the conflict over? Papa claims many royalists declared that they will not adhere to the new constitution because the prisoner King Louis was forced to sign it under duress. This means the hostility could escalate.

Sunday—September 18, 1791

We went to Notre Dame for a Mass attended by King Louis and Queen Marie Antoinette. The choir sang *Te Deum* to thank the heavens for inspiring the king to accept the new Constitution of France. Afterwards we went to the Champs Élysées to partake in the festivities. A hot-air balloon trailing tricolor ribbons glided over the city to announce the good news.

All is quiet now. I am alone in the house, waiting for my parents and my sister, who went to visit Marie-Madeleine. She had a baby boy two days ago. Madeleine and her husband named the baby Jacques-Amant. I stayed behind because I had a terrible headache and needed a nap.

I read a book that describes how, in a letter to Euler, Christian Goldbach conjectured that every odd number is the sum of three primes. When Goldbach wrote that letter, it was still not clear whether the number 1 should be considered a prime, but Goldbach apparently was assuming that it was. In my opinion, if 1 is excluded as a prime number, then Goldbach's conjecture should be rewritten as follows: "Every even number 4 and greater can be expressed as the sum of two primes, and every odd number 7 and greater can be expressed as the sum of three primes."

It looks simple. Yet, after many years mathematicians have not been able to prove Goldbach's conjecture! This is the type of problem that I would like to solve. But I need to know more about prime numbers. Perhaps if I were instructed by a mathematician I could learn faster.

I hear voices in the floor downstairs. It must be my sister and my parents returning from visiting Madeleine. I must go now to join them for dinner.

For my name day, Madeleine brought me a set of writing quills, and Angelique gave me a beautiful picture she sketched. With her charcoal drawing my sister captured the lively atmosphere of this part of Paris; it shows the Pont au Change, the Pont Notre Dame, and the Quai de la Grève in incredible detail.

Admiring the picture, I realized that Angelique has a keen visual sense, like an artist. In the background she sketched the Conciergerie and the Palais de Justice. She drew the piles of rubble that still remain near the bridge after the houses there were demolished. The most amazing thing is how Angelique caught the movement of people as if the lively scene she saw was frozen in time in her sketch. In the foreground she drew a girl in a white dress that resembles her. She also drew a lemonade vendor, a woman selling crêpes, one couple walking their dog, and people walking across the bridges. In her sketch the neighborhood looks charming and gay, just as I like to remember it. On the left corner of the picturesque scene, Angelique drew a group of National Guard soldiers marching with flags waving, heavily armed, a reminder that Paris is now besieged by a revolution. I will treasure my sister's gift forever.

I received more gifts. Millie gave me a lovely white handkerchief embroidered with my initials in red and blue. Madame Morrell fixed a delicious lunch of *pâté en croûte*, mushroom soup, and my favorite chocolate mousse cake.

Mother gave me the best gift of all: she told Papa that I should have a full-size desk in my bedroom. I was speechless. I wanted to hug her and thank her for granting me this wish. I always wanted a large desk, like the one in Papa's library. My lap desk is no longer adequate for writing my extensive notes. I don't know if this is a sign that Maman relented regarding my studies of mathematics. She did not say, and I did not ask her, but I kissed her cheeks tenderly to show my appreciation. Papa was also delighted with the idea, promising to order a desk from the English cabinetmaker, adding that I should have a bookcase to match, to contain my growing book collection.

I am so happy! I will place my desk near the window so that I can have a view of the sky to draw inspiration when I study.

Mother wanted a chance to get out of the city for some fresh air. She planned a picnic at the *Désert de Retz,* where we had a most enjoyable day. The *Désert* is located about 20 kilometers from Paris. Papa engaged a four-horse berline coach to carry the whole family, including Millie.

The *Désert de Retz* is a gorgeous and unusual garden, which was designed and constructed a few years ago by M. François Racine de Monville, a wealthy aristocrat who is an acquaintance of my father. The complex includes a botanical garden with rare and exotic varieties of trees, ornamental plants, and flowers imported from around the world.

Maman brought a basket of food for our picnic lunch that we ate under the trees. After lunch, Angelique went around looking for inspiration to draw landscape pictures, and I took a long stroll, alone, while my parents conversed. It is wonderful how the mind feels refreshed, walking among the trees, hearing only the sounds of birds singing and the tree leaves moving in the gentle breeze. I imagine that the great thinkers in history developed the most beautiful ideas while ambling in solitude, inspired by the sounds of nature.

Sunday—October 2, 1791

Today is the first day of a new era in the governing of France. The National Assembly dissolved. The deputies accomplished what they had set out to do two years ago. The new Constitution of France was presented to the nation, and the king accepted and signed it. Now that the Assembly's work is complete, a new government was elected.

Monsieur LeBlanc chatted with me yesterday. "Sophie," he said, "imagine I have five books, and I offer to lend you one, how many ways are there of selecting a single book?" The answer was easy, I said, "Five ways." He nodded and continued, "If we label the books A, B, C, D, and E, how many combinations can you make to select two books?" I listed all the book combinations: AB, AC, AD, AE, BC, BD, BE, CD, CE, DE. That makes ten selections. M. LeBlanc agreed with me. He did not give up and asked again, "What about if I lend you three from five?" Well, picking three books from five is the same as discarding two from five so there are ten ways of doing this.

M. LeBlanc was satisfied with my answers. Before leaving to join Papa, he mentioned that Antoine-August is studying this topic, which is called combinations. I stayed behind, trying to understand the relevance of mathematical combinations. I consulted a book that deals with the combination problem. It starts with the binomials. Just like the polynomial is an algebraic expression with multiple terms, a binomial is an algebraic expression with only two terms. The integral powers of the binomial are:

$$(x + y)^0 = 1$$
$$(x + y)^1 = 1x + 1y$$
$$(x + y)^2 = 1x^2 + 2xy + 1y^2$$
$$(x + y)^3 = 1x^3 + 3x^2y + 3xy^2 + 1y^3$$

If I place the coefficients (the numbers on their own and in front of the $x$'s) of the terms in the binomial expressions in a triangular array, with $(x + y)^0$ at the top, the coefficients of $(x + y)^1$ in the second row, those of $(x + y)^2$ in the third row, and so on, the resulting arithmetic array that results from this arrangement is called "Pascal's Triangle." It looks like this:

```
n = 0                        1
  = 1                    1       1
  = 2                1       2       1
  = 3            1       3       3       1
  = 4        1       4       6       4       1
  = 5    1       5       10      10      5       1
```

Pascal's Triangle is named after the French mathematician Blaise Pascal, who described it in his posthumously published book *Traité du triangle arithmétique*. It works for all positive whole number values of the power index. I can prove, for example, that,

$$(1 + x)^4 = 1 + 4x + 6x^2 + 4x^3 + x^4$$

Indeed, the coefficients of the binomial expansion with $n = 4$ are the 1, 4, 6, 4, 1 from Pascal's triangle.

Pascal's triangle has many properties and contains many interesting patterns of numbers. Some simple patterns I see by examining the diagonals in the triangle: (a) The diagonals going along the left and right edges contain only 1's. (b) The diagonals next to the edge diagonals contain the natural numbers in order. (c) Moving inwards, the next pair of diagonals contain the triangular numbers in order.

What if I colored the odd numbers on the triangle, would I get an interesting shape? I know that Pascal used the triangle to solve problems in probability theory. What else could I do with Pascal's Triangle?

Friday—October 7, 1791

$I$ will solve the following binomial: $(2 + 3x)^5$. Applying the general rule of ascending and descending indexes, and the coefficients from Pascal's Triangle, I can expand as follows:

$$(2 + 3x)^5 = 2^5 + 5(2^4)(3x) + 10(2^3)(3x)^2 + 10(2^2)(3x)^3 + 5(2)(3x)^4 + (3x)^5$$

This simplifies to

$$(2 + 3x)^5 = 32 + 240x + 720x^2 + 1080x^3 + 810x^4 + 243x^5$$

Algebraic expansions of binomial equations are related to the selection of combinations. So, if somebody asks me, "How many

150

ways are there of selecting six objects from eight?" I can check Pascal's Triangle. I see that, since there are total of eight objects, I look at the line beginning with 1, 8, etc. I want to select six so I count along from zero until I get to six. The number there is 28, so there are 28 selections.

Pascal's Triangle is very useful for finding the coefficients in the binomial expansion, especially if the power is $n < 6$. But it gets cumbersome when the exponent is greater, as one page is not big enough to contain all the rows of the triangle. For n > 7, it would be better to use a formula to calculate the number of combinations.

I am going to use Pascal's Triangle to play with Angelique. Maybe I can invent a game of combination of cards. This would be an interesting pastime to play with her during the quiet days of winter, when it becomes so cold that we stay indoors for weeks.

Saturday—October 15, 1791

Autumn is beautiful. During these months before winter the light of day seems a little softer, and the sunsets are more colorful. There is also some melancholic feeling in the air, or maybe is my own nostalgia, realizing that I'm leaving my childhood behind.

My little sister is also growing up. I was so absorbed in my studies that I did not notice the change in her. She traded her childish patter for a more contained countenance. She now moves with grace, without running wildly through the halls like a playful puppy. This evening she came to my bedroom and asked me, "Sophie, what is love?" I was taken aback by her question, for I had never thought about such matters. Angelique seemed flushed when she asked, and was eager for an answer. I did not know what to say, except to hold her hand and say that I love her.

But Angelique desired to know about love between two people who are not related. What *is* indeed that kind of love? I ask: Is love the emotion that my parents felt when they met, the passion that drew my sister Madeleine to her husband? If one pays attention, one sees in people certain tenderness when they are with a dear

151

beloved friend. That is the kind of love Angelique was curious to know.

After my sister left, I wondered more about her question, but even now I still do not have an answer. I only know mathematics. I could tell Angelique about *nombres amicaux*, amicable numbers. Greek mathematicians defined two whole numbers to be "amicable" if each was the sum of the proper divisors of the other. Reminiscent of a strong friendship between two people, a relationship that defines one person in terms of the other. The best way to illustrate amicable numbers is to give an example.

The pair 220 and 284 has the curious property that each number contains the other, in the sense that the sum of the proper positive divisors of each number sums the other. Let's see: the divisors of 220 are 1, 2, 4, 5, 10, 11, 20, 22, 44, 55, 110, and of course, 220 itself; but the last number is self evident, so I take only the other eleven divisors and add them up:

$$1 + 2 + 4 + 5 + 10 + 11 + 20 + 22 + 44 + 55 + 110 = 284$$

The divisors of 284 are 1, 2, 4, 71, 142, and when I add them up I get 220, that is:

$$1 + 2 + 4 + 71 + 142 = 220$$

That's what I mean saying 220 and 284 are amicable, as each number "contains" the other.

Amicable numbers were used in magic and astrology and in the making of love potions and talismans. The Pythagoreans knew of these numbers. They called 284 and 220 amicable numbers, adopting virtues and social qualities because the parts of each number have the power to generate the other. It is said that Pythagoras once said that "a friend is one who is the other I, such as are 220 and 284."

Are there other amicable numbers? Yes! The numbers 17,296 and 18, 416 were the next pair of amicables discovered by an Arabic mathematician named Ibn al-Banna in the thirteenth century. Then, in 1638, the French mathematician René Descartes found the more astonishing pair 9,363,584 and 9,437,056. I wonder how the mathematicians found amicable numbers. I haven't found a formula or method to uncover all of the amicable numbers. However, formulas for certain special types were discovered throughout the years.

An ancient mathematician named Thabit ibn Kurrah noted that if $n > 1$ and each of $p = 3 \cdot 2^{n-1} - 1$, $q = 3 \cdot 2^n - 1$, and $r = 9 \cdot 2^{2n-1} - 1$ are prime, and then $2^n pq$ and $2^n r$ are amicable numbers. In 1636, Fermat uncovered al-Banna's pair 17,296 and 18,416 using the formula with $n = 4$, as he reported in a letter to Mersenne. Two years later, Descartes wrote to Mersenne with the pair 9,363,584 and 9,437,056 for $n = 7$. By 1747, Euler had found thirty additional pairs, and then he added about thirty more new amicable pairs.

It's impressive. However, none of the books that mention these incredible accomplishments describe how the mathematicians discovered the amicable pairs. Surely, they did not take numbers and check their divisors one by one. This would not be mathematical analysis. There must be a method or an equation that generates amicable numbers.

Oh, but I digress. I started with a question of love and ended up with mathematics. I do not expect that Angelique would understand it.

Monday—November 14, 1791

I am still upset and exasperated. It's hard to put it in words, but I must attempt to write how insulted I feel by the words of a foolish man. I went with my parents on Saturday to the Theatre de la Nation to see *La Partie de Chasse de Henri IV*. I was enjoying the play, but an unfortunate encounter at the intermission ruined it all.

There I was with Maman and Papa, taking a refreshment, when M. and Mme Salvaterry approached us. Mme Salvaterry is a frivolous woman. She kept looking at her reflection in the hall's mirror, arranging her curls while talking nonsense. Her husband is not very nice either. I do not even recall how the conversation moved to my interest in mathematics; all I remember is that M. Salvaterry remarked, without even looking at me, "Ah Monsieur Germain, your clever daughter must read Francesco Algarotti's books. They are simple enough, so maybe she can understand such abstract matters."

I was furious. How dare he suggest I read such inane books? And then his wife added with a forced smile, "They are wonderful, my dear!" and went back to rearranging her hair as if that was all she knew how to do. My father was about to say something but was interrupted by the call for the second act, and we returned to our seats. I am upset that anybody would suggest I read such nonsense.

That night I lay in bed, listening to the wind outside my window, unable to sleep, asking why anyone would think that I'm stupid to read Francesco Algarotti's books. He wrote "Science Books for the Use of Ladies," believing that women are only interested in romance. So he attempts to explain science and mathematics through the flirtatious dialogue between two silly people in love. Those books are girlish, total nonsense! Oh Papa, why you did not say that I am studying to become a true scholar? I only read books written by real mathematicians.

How I wish I could articulate my thoughts and speak my mind. I would have said that I am not interested in frivolities. Alas, I do not know how to say these things without sounding ill-mannered.

Tuesday—November 29, 1791

I have amused myself with binomial expansions, fascinated by their symmetry and how easy it is to find the coefficients. Papa told me that binomials are used in gambling games, and then he told me a little bit about how this part of mathematics was developed.

To begin with, although the arithmetic triangle is named after the French mathematician Blaise Pascal, several other mathematicians knew about and applied their knowledge of the triangle centuries before Pascal was even born. The arithmetical triangle was discovered independently by both the Persians and the Chinese during the eleventh century. Chinese mathematician Chia Hsien used the triangle to extract square and cube roots of numbers. After Chia Hsien's discovery of the relationship between extracting roots and the binomial coefficients of the triangle, the Chinese continued work on this topic for many years in their effort to solve higher order equations.

The arithmetic triangle came to be known as Pascal's Triangle after Blaise Pascal published his *Traité du triangle arithmétique* (*Treatise on the Arithmetical Triangle*) in 1654. Pascal developed many of the triangle's properties and applications, making use of the already known array of binomial coefficients.

Pascal's work on the triangle stemmed from the popularity of gambling. A French nobleman had approached him with a question about gambling with dice. Then Pascal shared the question with Fermat, and after some thinking, the idea of the Arithmetical Triangle was born.

Using Pascal's Triangle, one can find the number of ways of choosing $k$ items from a set of $n$ items simply by looking at the $k^{th}$ entry on the $n^{th}$ row of the triangle. So, to see how many different trios one could form using the 45 members of a group, one would look at the third entry on the $45^{th}$ row. The 1 at the top of the triangle is considered the $0^{th}$ row, and the first entry on each row is labeled the $0^{th}$ entry on the row. The numbers in each row coincide with the coefficients of the binomial $(1 + x)^{n}$.

Sunday—December 25, 1791

Christmas is no longer a joyful celebration. We had our traditional dinner and went to church, but there was no *Pastorele*. I don't know if the new government had anything to do with it, or if the priests were uneasy to engage in a religious festivity. Maybe the people themselves are afraid.

I read about Galileo Galilei, a great Italian philosopher and scholar born in 1564. Galileo prided himself on having been the first to observe the sky through a telescope. He believed that his greatest quality was his ability to observe the world at hand to understand the behaviour of its components, and to describe them in terms of mathematical proportions. In fact, Galileo measured the swinging of pendulums until he could describe their periods by a mathematical law. He rolled bronze balls down inclined planes to describe the rate of acceleration in free fall. He performed many experiments, and he deduced the mathematical equations that govern the motion of many things in nature.

In the seventeenth century, the sun and all the planets were thought to revolve around the Earth. When Galileo learned the new heliocentric theory proposed by Nicolaus Copernicus, that the Earth and all the other planets revolve around the sun, he embraced it and became its best proponent. His observations with his new telescope convinced him that indeed the planets revolve around the sun. But accepting the heliocentric theory caused Galileo much trouble with the Roman Church.

In the spring of 1633, Galileo appeared before the Roman Inquisition to be tried on charges of heresy. He was denounced, according to a formal statement, "for holding as true the false doctrine that the sun is the center of the world, and immovable, and that the earth moves!" The great scholar was found guilty and forced to renounce his views. The Inquisition sentenced Galileo to a life of perpetual imprisonment and penance. Due to his advanced age, the Inquisition allowed him to serve the sentence under house arrest at his villa in Arcetri in Florence.

I sit in front of my window, looking at the sky full of brilliant stars shimmering in the darkness of the cold night. The feeble flame of my candle trembles, casting shadows on my writing. My mind wanders. Reading about Galileo confirms my idea of the marvelous intricate connection between mathematics and the universe. I, too, wish to formulate a mathematical law that describes a physical phenomenon. But learning about Galileo's last days also makes me realize that religious beliefs can be in conflict with an intellectual idea. How can a scholar compromise both?

Saturday—December 31, 1791

I wrote an essay to summarize my recollection of what happened in 1791, and to reflect on the social changes and the new government. I asked Angelique to draw a picture of Paris to illustrate my writing, to leave a permanent record of what time might erase from our memory.

Much happened in the past twelve months. Reading my journal notes, I conclude that 1791 was the year of confusion and violence. King Louis fled Paris, was captured, and returned to Paris. The people were against the king, then supported him, then recognized his power, and then they did not. King Louis first ignored the new constitution, and then he accepted it, although he did it under duress.

The government also seemed confused. First the deputies supported the king and restored his prerogatives, and then they suspected him. The Constituent Assembly dissolved amidst conflicts and dissentions. The Legislative Assembly was formed. Violence erupted throughout 1791, leaving in my memory the horrible image of the massacres of the Champ de Mars, when many people were killed during the protest against the reinstatement of King Louis XVI.

It is almost midnight, just a few minutes before a new year begins. I wonder, what does the future hold for the people of France? Will the violence and horrors leave way for reconciliation? If I could have a wish granted, I'd only ask for peace.

☾✸☽

Angry mob attempts demolition of Vincennes castle - February 28, 1791.

Arrival of royal family in Paris after their failed attempt to flee, June 25, 1791.

*Paris, France*
*Year 1792*

# 4

## Under Siege

I BEGIN THE NEW YEAR with more determination and a renewed resolve to study prime numbers. One of my goals is to acquire the necessary mathematical background to prove theorems.

Prime numbers are exquisite. They are whole pure numbers, and I can manipulate them in myriad ways, as pieces on the chessboard. Not all moves are correct but the right ones make you win. Take, for example, the process to uncover primes from whole numbers. Starting with the realization that any whole number $n$ belongs to one of four different categories:

The number is an exact multiple of 4:  $n = 4k$
The number is one more than a multiple of 4:  $n = 4k + 1$
The number is two more than a multiple of 4:  $n = 4k + 2$
The number is three more than a multiple of 4:  $n = 4k + 3$

It is easy to verify that the first and third categories yield only even numbers greater than 4. For example for any number such as $k = 3, 5, 6$, and 7, I write: $n = 4(3) = 12$, and $n = 4(6) = 24$; or $n = 4(5) + 2 = 22$, and $n = 4(7) + 2 = 30$. The resulting numbers clearly are not primes. Thus, I can categorically say that prime numbers cannot be written as $n = 4k$, or $n = 4k + 2$. That leaves the other two categories.

So, a prime number greater than 2 can be written as either $n = 4k + 1$, or $n = 4k + 3$. For example, for $k = 1$ it yields $n = 4(1) + 1 = 5$, and $n = 4(1) + 3 = 7$, both are indeed primes. Does this apply for any $k$? Can I find primes by using this relation? Take another value such as $k = 11$, so $n = 4(11) + 1 = 45$, and $n = 4(11) + 3$. Are 45 and 47 prime numbers? Well, I know 47 is a prime number, but 45 is not because it is a whole number that can be written as the product of 9 and 5. So, the relation $n = 4k + 1$ will not produce prime numbers all the time.

Over a hundred years ago Pierre de Fermat concluded that "odd numbers of the form $n = 4k + 3$ cannot be written as a sum of two perfect squares." He asserted simply that $n = 4k + 3 \neq a^2 + b^2$. For example, for $k = 6$, $n = 4(6) + 3 = 27$, clearly 27 cannot be written as the sum of two perfect squares. And of course, I can verify this with any other value of $k$.

Sunday—January 8, 1792

It's bitterly cold tonight. I hear the wheels of carriages crunching the ice on the streets, sounding as glass breaking. Before going to bed Millie brought me a cup of hot chocolate; she left me a tall candle and restocked my heater. It's after midnight, but I cannot go to sleep just yet.

I found the way to solve a problem that had vexed me for days. It just came to me, so clear, so simple. I know just what to do to prove Fermat's assertion that odd numbers of the form *n = 4k + 3* cannot be written as a sum of two perfect squares $a^2 + b^2$.

I consider three cases:

Case 1: If both $a$ and $b$ are odd numbers, their squares $a^2$ and $b^2$ are odd as well. Therefore, the sum of two odd numbers will be even.

For example $a = 3$ and $b = 5$, so that $a^2 = 9$ and $b^2 = 25$, thus $a^2 + b^2 = 9 + 25 = 34$.

Therefore, I conclude that the odd number $n = 4k + 3$ cannot be written as a sum of these perfect squares $a^2 + b^2$ when $a$ and $b$ are odd numbers.

Case 2: If both $a$ and $b$ are even numbers, their squares $a^2$ and $b^2$ are even as well. Therefore, the sum of two even numbers will be even.

For example $a = 2$ and $b = 4$, so that $a^2 = 4$ and $b^2 = 16$, thus $a^2 + b^2$ $= 4 + 16 = 20$.

Therefore, I conclude that the odd number $n = 4k + 3$ cannot be written as a sum of these perfect squares $a^2 + b^2$ when $a$ and $b$ are even numbers.

<u>Case 3</u>: I consider the sum of one square of an even number and one square of an odd number. If $a$ is an even number, it means that it is a multiple of 2, thus $a = 2m$ where $m$ is a whole number. On the other hand, I assume $b$ is an odd number, and as such it is one more than a multiple of 2, or $b = 2r + 1$, where $r$ is a whole number. Therefore,

$$a^2 + b^2 = (2m)^2 + (2r + 1)^2$$
$$= 4m^2 + (4r^2 + 4r + 1)$$
$$= 4(m^2 + r^2 + r) + 1$$

This means that $a^2 + b^2$ is one more than $4(m^2 + r^2 + r)$. However, it cannot be $n = 4k + 3$ since this is three more than a multiple of 4.

This proves that an odd number of the form $n = 4k + 3$ cannot be written as the sum of two perfect squares $a^2 + b^2$, regardless of whether the two squares are odd or even, or a combination of the two.

Q.E.D.

I did it! Now I can go to sleep.


Thursday—January 12, 1792

When I get discouraged, trying in vain to solve a difficult problem, I have to remind myself that other people also struggle. It inspires me and helps me not to give up. I've learned that one must be fully devoted to study mathematics, and sometimes one has to persevere against all odds. Many mathematicians probably had the privilege to pursue their studies in great comfort, having tutors to teach and guide them. However, many others had to make a great effort to overcome not only mathematical challenges, but life obstacles as well.

Jean Le Rond D'Alembert, for example, was abandoned by his mother when he was a baby. She left him on the steps of a church in Paris because he was an illegitimate child. The priests found the baby and gave him to family to raise him. The adopting parents sent the boy to school where his tutors discovered his genius in mathematics. D'Alembert overcame his miserable beginnings and became one of the greatest French mathematicians of this century.

Another example is Leonhard Euler, a mathematician that faced a physical challenge when he became blind. Yet, losing his sight did not stop his mathematical creativity and genius. And then there was his friend Daniel Bernoulli, who faced opposition from his father to become a mathematician. Bernoulli loved mathematics since he was a child, but his father ordered him to become a merchant and marry a girl he did not like. Once it became clear that Daniel was not good in business, his father then forced him to become a physician. Daniel Bernoulli did study medicine, but eventually he became a great mathematician.

I can relate to Daniel Bernoulli's struggle. It confirms my belief that if one wants to become mathematician, nothing can prevent it. Mathematics is an abstraction of the mind, an intellectual desire that cannot be stifled. However, mathematics can't be grasped by a mere glimpse, by simply reading books; it requires serious study and perseverance. I must devote my heart, soul, and every bit of my mental energy to mathematics if I ever wish to master it.

Wednesday—February 1, 1792

Mother is not well. I am worried because her chest pains and headaches are getting more severe. She is always nervous, fretting about Papa; when he is late, she fears he was trampled by the bloodthirsty mobs that roam the streets. Madame de Maillard comes daily to help nurse Mother back to health.

I am fond of Mme de Maillard because she is the only lady who understands my desire to study. She is an intelligent lady, not afraid to express her opinions about the ideals of the revolution. Madame reads the works of great philosophers and can debate anybody with excellent arguments. She says that women are endowed by nature with the same mental capabilities as men. However, Madame also believes that women have a natural duty to remain in the domestic sphere to raise children. She argues that women should be educated in order to rear children intelligently and guide them in their studies. I agree with her about the right to education, especially if she means a right to pursue scientific studies, which is only a privilege of men; but I believe that women should choose whether to raise children or not. I cannot agree on forcing upon women roles they do not wish to play.

I hope I am not catching anything. The chill of winter lingers in my body. My hands are always cold, my legs ache, and I have a cough. During the day I study near the fireplace in my father's library to stay warm. I wear two pairs of wool stockings to bed, and Millie has to warm my blankets with the hot iron. But even if it is cold, I prefer to work at night, when the only sound in the house is the chiming of the clock announcing each hour.

The night is mine to lose myself in the secrets of mathematics. When I master a new topic I feel excited, and my dreams carry me to new dimensions. When I go to sleep without solving a problem or am unable to understand a new concept, dreams do not come easy, just like tonight. It is cold, and my body aches, but my mind is restless, thinking and thinking about numbers.

Friday—February 24, 1792

The snow covers the bruises of the city. Yes, Paris is wounded, tormented by the ravages of social chaos, bloodshed, and economic troubles. The winter is harsh, aggravating the devastation and the misery of the poorest people. It's a difficult time for France now.

Millie complains how hard it is to find groceries. Robberies make it more difficult for the shop owners. Scamps run off with stolen goods and sell them at the other end of the street. Today, Millie caught two thieves loaded with sugar and coffee running along the Rue des Lombards. The shoplifters approached her and tried to sell her the sugar at exorbitant prices. Even candles are scarce, and oil for the lamps is very expensive. I don't want to aggravate the situation so I've given up studying at night to save my candles.

My mother is slowly recovering from her illness. Her poor health, the gray bitter winter, and the tumultuous political revolution that goes on in Paris have contributed to the collective depression that surrounds us. Even Millie, usually so vivacious, comes home from her errands not only tired and chilled but also depressed. Without Maman's musical soirées to entertain us every week, the house feels colder and darker than usual. If it wasn't for Angelique's chattering enthusiasm, we'd be miserable.

Saturday—March 3, 1792

It was a lovely party. Some of our friends joined my family this evening to partake in a musical interlude planned especially for Maman. A few weeks ago Angelique came up with the idea of doing something to cheer her up. I suggested staging the play *Esther* by Racine because it's a favorite of my mother's. This play tells the story of Haman and the Jewish queen Esther who risks her own life to save her people from certain destruction.

Angelique got Millie all excited about performing in the play. She wanted to give me a part as well but I refused, proposing instead to help them in other ways. My sister chose to play Esther and gave Millie the part of King Ahasuerus. With Madame Morrell's assistance, they made the costumes and I helped Angelique and Millie memorize their lines and practice their dialogue. I volunteered to play the musical choral interludes at the piano because there was no one else to do it.

166

The play was a success. Angelique invited Madeleine, her husband, and M. and Mme de Maillard. Maman did not know about the play, thus she was pleasantly surprised when Mme Morrell opened the drapes to reveal the improvised stage she and her husband helped my sister to decorate. Maman was ecstatic, and I noticed a few tears in her eyes during the monologue when my sister recited the words, "Oh God, my king, behold me trembling and alone before thee!" She was convincing as Esther. Maman was so pleased; at the end of the performance she broke out in applause along with the others.

Tuesday—March 13, 1792

I discovered a most peculiar book. The first page starts with the following statement:

*"A certain man put a pair of rabbits in a cage. How many pairs of rabbits can be produced from that pair in a year if it is supposed that every month each pair begets a new pair which from the second month on becomes productive?"*

It was written hundreds of years ago by a mathematician known as Fibonacci. At first I thought it odd that a book written by a mathematician should deal with such infantile problem. But soon the puzzle caught my interest, and I decided to learn more about it. Leonardo Pisano, better known by his nickname Fibonacci, published in 1202 a book titled *Liber Abaci* (*Book of the Abacus*), based on the arithmetic and algebra that he learned in his trips to far away countries.

Fibonacci introduced the Hindu-Arabic decimal system and the use of Arabic numerals into Europe. His book also included simultaneous linear equations. The second section of *Liber Abaci* contained a large collection of problems aimed at merchants. They relate to the price of goods, describing how to calculate profit on transactions, how to convert between the various currencies in use in Mediterranean countries, and many other practical problems.

The problem of the rabbits appeared in the third section of *Liber Abaci*. The original problem that Fibonacci investigated was, simply stated, how fast rabbits could breed. At first the problem seems inconsequential, but it requires logical thinking to solve it. This is the way I analyze the problem:

Suppose the man starts with a newly born pair of rabbits, one male and one female, and he places them in a cage. Rabbits are able to mate at the age of one month, so at the end of the second month a female can produce another pair of rabbits. Suppose that the rabbits never die and that the female always produces one new pair (one male, one female) every month from the second month on. Fibonacci asked, "How many pairs will be in one year?" Well, I will try to answer:

1. At the end of the first month, the rabbits mate. There is now 1 pair.

2. At the end of the second month the female produces a new pair, so now there are 2 pairs of rabbits in the cage.

3. At the end of the third month, the original female produces a second pair, making 3 pairs in all in the cage.

4. At the end of the fourth month, the original female has produced yet another new pair, the female born two months ago produces her first pair also, so there are now 5 pairs in the cage.

The number of pairs of rabbits in the cage at the start of each month is 1, 1, 2, 3, 5, 8, 13, 21, 34, 55, 89, 144. These numbers are called Fibonacci's numbers. And the answer to the original question is 144; that is, there will be 144 rabbits in the cage at the end of 12 months.

This problem is not very realistic, is it? It implies that sibling rabbits can mate. I would rather say that the female rabbit of each pair mates with any non-brother male and produces another pair. Also, the problem implies that each birth is of exactly two rabbits, one male and one female. This is not true in life. Perhaps Fibonacci used rabbits as an illustration in the sense of mathematical objects rather than as living creatures. He made his point this way. He used fictitious animals to illustrate other problems, such as, "A spider climbs so many feet up a wall each day and slips back a fixed number each night, how many days does the spider take to climb the wall?" And, "A hound whose speed increases arithmetically chases a hare whose speed also increases arithmetically, how far do they travel before the hound catches the hare?"

How important are Fibonacci's numbers in mathematics? Or are they only a game, an intellectual diversion? How many Fibonacci's numbers are there? Infinite?

I know how to form the Fibonacci's series. I start with 0 and 1 and then add the latest two numbers to get the next one as follows:

the series starts like this:    0 1

$$0 + 1 = 1$$

so the series continues...    0 1 1

and the next term is    $1 + 1 = 2$

so we now have    0 1  1 2

$$1 + 2 = 3$$

then    0 1 1 2 3

and the series continues.....

$n$:    0  1  2  3  4  5  6  7  8  9  10  11  12  13  14  15 16 ...

Fib($n$): 0  1  1  2  3  5  8  13  21  34  55  89 144 233 377 610 987...

*Très bien*. Now that I understand what it is, what do I do with Fibonacci's series?

Wednesday—March 21, 1792

$W$hat a glorious day! Although there is still a chill in the air, the sun came out, and the streets were full of people. Since waking up, I felt euphoric, in a cheery mood. After dinner I had a long conversation with Papa. I asked him if he would take me one day to attend the lectures at the Academy of Sciences. I know that every week mathematicians present their work in the king's hall. It would be the most exciting experience of my life to listen to these great men talk about their scholarly work. I am so curious to hear how they explain mathematics and learn how they apply the analysis to the study nature. Even though the lectures are public, one needs special tickets, which are reserved for colleagues and important visitors. I'd give anything to gain admittance to the Academy lectures.

I want to meet any scholar, but there is one in particular that seems very interesting. His name is Joseph-Louis Lagrange. He is originally from Italy, but Lagrange worked in Germany and, after being recommended by French mathematicians, the king invited and offered M. Lagrange a post in the Paris Academy of Sciences. Father knows someone who met him and claims that Lagrange is a very nice, soft-spoken, and humble individual. And because M. Lagrange is a great mathematician, my dream is to meet him in person one day.

Palm Sunday—April 1, 1792

To celebrate my sixteenth birthday I had a *tête-à-tête* with Papa. We talked about my studies and his hope for a better future. I've been gloomy, and to cheer me up Papa gave me a lovely fan imported from England. But a material gift can't cure my melancholy.

I also miss Angelique's chatter. My sister has influenza and has stayed in bed for over a week. I cannot visit her because Maman is afraid her illness is contagious. I've spent the time alone, reading and translating the last chapters of Euler's *Introduction to Analysis*. It took me a very long time, but I am almost finished.

Mme de Maillard gave me the *Pensées* (*Thoughts*), a book written by Blaise Pascal, the same French mathematician who studied the arithmetic triangle. The *Pensées* is a collection of hundreds of notes Pascal made, many of them intended for a book, which he expected would help him defend Christianity. Pascal organized and classified some notes before he died, but others remain unsorted, and the book he intended was never written. Pascal's *Pensées* are powerfully insightful thoughts, which lead us more deeply into contemplation of human nature and the strivings of the heart and mind.

Traditionally on this day we start the celebration of Holy Week with a special Mass at Notre Dame. But under the new government rules it was discouraged. The churches are now almost empty; the faithful, like my mother, refuse to take communion from the juror priests, and the rest of the people are angry with the clergy. How fitting, I spent my birthday reading the philosophy of a mathematician who defended the Church.

Easter Sunday—April 8, 1792

$P$erfect numbers have fascinated mathematicians since ancient times. Many scholars were concerned with the relationship of a number and the sum of its divisors, often giving mystic interpretations. A positive integer $n$ is called a perfect number if it is equal to the sum of all of its positive divisors, excluding $n$ itself.

Perfect numbers have the form $2^{n-1}(2^n - 1)$. To see how this works, I start with the first prime number $n = 2$, and $(2^n - 1) = 3$, which is also prime; therefore:

$$2^{n-1}(2^n - 1) = 2^{2-1}(2^2 - 1) = 2^1(4 - 1) = 2\,(3) = 6$$

This shows that 6 is the first perfect number because $6 = 1 + 2 + 3$. The next number is 28 since $1 + 2 + 4 + 7 + 14 = 28$. The next two perfect numbers are 496 and 8128. These four perfect numbers were all known before the time of Christ. More than two thousand years ago, the Greek mathematician Euclid explained a method for finding perfect numbers that is based on the concept of prime numbers.

There are two theorems related to perfect numbers:

Theorem I: $k$ is an even perfect number if and only if it has the form $2^{n-1}(2^n - 1)$, and $(2^n - 1)$ is also prime.

Theorem II: If $(2^n - 1)$ is prime, then so is $n$.

171

The first perfect numbers 6, 28, 495, and 8128 are even numbers. Are there odd perfect numbers? I could try to verify this by applying the formula $2^{n-1}(2^n - 1)$ to each prime number one by one. However, this is an enormous task, considering that the fourth perfect number is already a large number; it would probably take all my life. Besides, this is not the way to prove a mathematical statement is true (or not).

I would like to prove that there are infinite perfect numbers. I am sure it would require first proving that there are infinite prime numbers. For the time being, I conclude that even perfect numbers have the form $2^{n-1}(2^n - 1)$, where the number $(2^n - 1)$ is itself a prime number, and the exponent $n$ is also prime.

How I wish I could share my analysis with a mathematician or someone who understands. If I show it to Papa he will be impressed, but he will not know what to make of it.


Friday—April 27, 1792


$A$ncient Greek mathematicians studied prime numbers and their properties. Euclid's *Elements* contains several important results about primes. In Book IX of the *Elements*, Euclid proved that there are infinitely many prime numbers. This proof uses the method of contradiction to establish a result. Euclid showed that if the number $(2^n - 1)$ is prime, then the number $2^{n-1}(2^n - 1)$ is a perfect number.

In the year 200 B.C., Eratosthenes devised an algorithm for calculating primes called the Sieve of Eratosthenes. Eratosthenes' method is so named because it sifts out multiples of numbers from the set of natural numbers (excluding 1), so only the prime numbers remain. Then there is a long gap in the history of prime numbers that lasted several hundred years until 1651, when the French mathematician Pierre de Fermat revived the study of prime numbers. He corresponded with other mathematicians, in particularly with Marin Mersenne, who was trying to find a formula that would represent all primes. Although he failed in this, Mersenne's work on numbers of the form $2^p - 1$, where $p$ is a prime number, helps in the investigation of large primes.

In one letter to Mersenne, Fermat conjectured that the numbers $2^n + 1$ were always prime if $n$ is a power of 2. Fermat verified this formula for $n = 1, 2, 4, 8$, and 16.

Mersenne also studied numbers of the form $2^n - 1$, discovering that not all numbers of this form, with $n$ prime, are prime numbers. For example, $2^{11} - 1 = 2047 = (23) \cdot (89)$ is a composite number. For many years numbers of this form provided the largest known primes. Extraordinary, there is so much to discover!

<br>

Tuesday—May 15, 1792

$I$s there a God? Is there an intelligent Being that created the universe and everything in it? I am consumed with questions of this type but cannot answer them. Sometimes I am not sure what to believe. The problem is that I doubt so much. Ever since I was a child I have prayed, as Mother taught me. But as I grew up I began to question it and to think it futile to pray.

Maman has always said that God answers our prayers. So, when I was a little girl I prayed to God to make me as lovely as my sister Madeleine. But every time I looked in the mirror, as Maman fixed my hair, I did not see any change. I also wondered about the unfortunate beggars sitting at the church steps; many of them are blind, lame, and diseased. Why would God allow such suffering and misery? When Maman was ill I prayed that she'd get better, and when she did I was not sure if she was cured because we prayed, or because she took the tonics the doctor prescribed. Once I even considered becoming a nun, but then I realized that it would be hypocritical since I am not sure of my faith.

My mother and sisters believe in a God that is omnipotent. They seem to find comfort praying to a God that no one sees, but they believe He is everywhere. We have witnessed the desolation and the terror that has devastated our country in the last few years.

We have prayed for others, whether they are innocent or guilty. We have prayed, but I am not sure it helps.

I read books written by French intellectuals who upheld the principles of Liberalism. Basically they say in their writings that man is responsible to no authority; that men owe nothing to God; and that the mind and will of man replaces the will of God. These three tenets were discussed many times in Father's library. Ever since I was a little girl, sitting quietly and listening to the adults talking, I tried to understand what all that means. I never see Father praying and wonder whether he, like me, feels any doubt when Mother says that God will save us all. It is hard to believe it after we witness the bloodshed and injustices all around us. One thing is for sure. I do hope Mother's prayers are answered and that the violence in the streets of Paris soon will end.

Wednesday—May 30, 1792

Papa gave me a beautiful book written by Leonhard Euler. It is titled *Lettres a une princess d'Allemagne sur divers sujets de physique & de philosophie* (*Letters to a German Princess on Different Subjects in Physics and Philosophy*). Papa explained that Euler wrote in this book the topics he taught to a girl like me. During his stay in Berlin, Euler was asked to provide tuition for Princess d'Anhalt Dessau, a niece of Frederick the Great. Euler instructed the princess through these letters.

At the start of their correspondence, Euler's royal pupil was a fairly ignorant girl. The princess did not know natural philosophy or mathematics. Thus, Euler began to teach her starting with the basic notions of distance, time, and velocity. The letters continued with more difficult aspects of physics: light and colour, sound, gravity, electricity, and magnetism. Euler then addressed the nature of matter and the origin of forces. The instruction of the German princess was extensive.

The *Letters* also cover metaphysics, dealing with the mind-body problem, free will and determinism, the nature of spirits, and the operation of Providence in the world of nature. Euler believed that, unaided, human reason will not take us very far, and that some philosophical questions will be left unanswered awaiting divine illumination.

I will study this book as diligently as the princess must have done it. I will pretend that each letter is addressed to me and imagine Euler patiently writing the concepts that he wishes me to learn.

Tuesday—June 5, 1792

Last night we went to the opera. Mother made me wear a new blue silk dress that was most uncomfortable. The full fichu of sheer organza made my neck itch. Maman insisted that I wear a corset too, so I had to sit uneasily upright through the evening. At least I did not have to wear the enormous pannier skirts that women wore years ago. Millie fixed my hair and, against my protests, Angelique put a little rouge on my cheeks. I am glad that they didn't powder my hair and pile it so high that it'd look grotesque. Papa gave me a very pretty fan for my birthday in April, so I took it with me to the concert. Angelique looked all grown up with her white gown and tricolor sash around her waist. She wore her hair curled just like Madame Royale, the young daughter of King Louis.

The opera *Les Evénemens Imprévus* had Mme Dugazon playing the *soubrette*, maid. We sat not far from the royal box, and Mother was very excited, looking forward to seeing the beautiful queen. Marie Antoinette soon showed up accompanied by the Dauphin and her daughter. Madame Elisabeth, the king's sister, and Madame Tourzelle, the governess, were also with her. I noticed from the beginning that Her Majesty seemed distressed. She was overwhelmed by the applause of her loyal subjects, and I saw her dab the tears from her eyes. The little Dauphin, who sat on the queen's lap, seemed anxious to know the cause of his mother's weeping.

In one act of the opera, Madame Dugazon sings a duet with the valet, and when she sang, "*Ah! Comme j'aime ma maîtresse*" ("Oh! How I love my mistress"), she looked directly at Marie Antoinette. At that moment a few angry citizens jumped upon the

175

stage, attempting to hurt Mme Dugazon. The other actors had to restrain the ruffians. During the commotion, the queen and her family left in a hurry. Maman was mortified and upset to see more and more people who so openly offend the queen. How can they show their scorn so blatantly?

There is a full moon tonight. I will snuff out my candle and sit by my bedroom window to watch the dark sky full of stars. Maybe I will recognize some constellations. In a night like this I feel drawn by the mysteries of the universe.

Sunday—June 10, 1792

$\Gamma$he numbers $\pi$ and $e$ occur everywhere in mathematics. Some time ago I proved analytically that $\pi$ is irrational using the method *reductio ad absurdum*. I first assumed that $\pi$ is represented by some ratio of whole numbers $p/q$. Since I could not find a fraction $p/q$ equal to $\pi$, I concluded that no fraction exists equal to $\pi$, and therefore it is an irrational number. The proof was relatively straightforward.

Then I tried to prove that Euler's number $e$ is also irrational, but I could not do it. I realize now that first I needed to establish some mathematical truths in order to make the appropriate assumptions. I think now I am ready. To prove that $e$ is irrational, I begin by stating a simple theorem:

Theorem:    $e$ is irrational.
Proof:       I know that for any integer $n$

$$e = 1 + \frac{1}{1!} + \frac{1}{2!} + \frac{1}{3!} + \dots + \frac{1}{n!} + R_n$$

where

$$0 < R_n < \frac{3}{(n+1)!}$$

First I assume that $e$ is rational, i.e. that there are two positive integers $a$ and $b$ such that $e = a / b$, and I let $n > b$. Then I rewrite the series as

$$\frac{a}{b} = 1 + \frac{1}{1!} + \frac{1}{2!} + \frac{1}{3!} + ... + \frac{1}{n!} + R_n$$

and, multiplying both sides by $n$-factorial, or $n!$

$$n!\frac{a}{b} = n! + \frac{n!}{1!} + \frac{n!}{2!} + \frac{n!}{3!} + ... + 1 + n!R_n$$

Since $n > b$, the left side of this equation represents an integer, therefore $n! R_n$ is also an integer. But I know that

$$0 < n!R_n < \frac{3n!}{(n+1)!} = \frac{3}{n+1}$$

So, if $n$ is large enough, the left side is less than 1. But then $n! R_n$ must be a positive integer less than 1, which is a contradiction.

Therefore, $e$ is an irrational number.
Q.E.D.

Thursday—June 21, 1792

A gang of angry sans-culottes invaded the Tuileries Palace yesterday. I saw the crowd ready for the attack at around one o'clock. My mother, Angelique and I were coming from visiting Mme LeBlanc when a group of sans-culottes blocked our carriage. Led by Santerre, the armed men were near the royal palace, rowdily shouting insults, and singing the dreaded song *Ça Ira*. The awful lyrics, *tous les aristocrats on les pendra*, "all the aristocrats will hang," resonated in my ears. It sounded so menacing that instinctively we knew something dreadful was about to happen.

I tried to forget the incident, but later Papa described in vivid detail what happened. The sans-culottes assaulted the royal palace, brandishing pistols and sabers in the king's face, shouting insults, screaming atrocities. The men humiliated the king and queen and threatened them with violence for several hours. They forced King Louis to don the bonnet *rouge*, the liberty cap, and to drink to the health of the people. The sans-culottes showed no respect and treated the king as if he was a puppet, addressing him as "monsieur" rather than as "His Majesty." The people of France have lost respect for King Louis. Gone are the days when he was treated with the highest reverence. This attack may be the beginning of the end of the reign of Louis XVI.

The social tension escalates. There is confusion regarding the law on the deportation of refractory priests, the king's veto of this law, the dismantling of the royal Constitutional Garde, and the stationing of thousand of troops by the Assembly in Paris. Last week Louis XVI dismissed his Jacobin ministers and replaced them with more moderate Feuillants. Amidst all this, the violence in the streets escalates. After what I witnessed, I am afraid the monarchy in France is over.

Saturday—June 30, 1792

It's terribly hot today. I wish we could go on vacation to our country home in Lisieux. I wish we could get away from the turbulence that rolls over Paris. But traveling is highly restricted; people need passports to leave the city, and the situation in the village is unstable.

For the same reason we hardly go to the Tuileries gardens or any other public place, as it is not possible to take a leisurely walk for fear of the mobs. My sister Angelique is getting more miserable sitting in the house all day. I suggested to Papa to take us to an art gallery for my sister's sake. Admiring the works of art will help her with her painting and ease her restlessness. She was thrilled with my suggestion and kissed my cheeks several times.

The only other outing for Angelique is to the dress shops in the Rue des Petits-Champs with Mother. Angelique is also making pretty rosettes from tricolor satin ribbons to pin to our dresses and hats. She made tricolor cockades for Papa's hats to show his support for the revolution.

What lies beyond the firmament? What alien worlds gravitate around other suns so far that I cannot see? A star near the edge of the night sky beckons me, urges me to seek. What should I search for? What is my destiny?

Perhaps it is the warm night or the quietness around me, but at this moment I do not wish to read or solve equations. Tonight I feel like looking up, far away, trying to decipher my destiny in the vast universe open before me.

Saturday—July 14, 1792

It's the second anniversary of the destruction of the Bastille. This time we did not celebrate. The violence and attempts against the king have turned Paris into a city of fear. Last week, the revolutionary government proclaimed that the *patrie est en danger.* Yes, we all are in danger, especially the royal family.

Paris resembles an armed camp. The black flag is flying over the Hôtel de Ville. Every day we see groups of soldiers parading in public places, heavily armed and chanting the *Ça Ira*, making me feel uneasy. It is like observing a dark sky full of clouds, foretelling a violent storm ready to erupt.

At home sometimes the tension makes us snap at each other. The other day Millie was cleaning Mother's bedroom, singing a catchy song, but Madame Morrell angrily asked her to stop. The chorus goes: "We will win, we will win, we will win. The people of this day never endingly sing. We will win, we will win, we will win. In spite of the traitors, all will succeed." Then I listened to another verse in the song: "Let's string up the aristocrats on the lampposts!"

179

It became clear. Millie was singing *Ça Ira*, a revolutionary song favorite among the sans-culottes to insult clergy, aristocrats, and the wealthy bourgeoisie. I am sure she hears the song in the streets so often that she sang it without even being aware what it means.

Monday—July 30, 1792

The soldiers from Marseille arrived in Paris this afternoon. Millie burst in out of breath to tell us. We raced to the open windows to catch the battalions of soldiers marching in, singing with loud tenor voices a patriotic hymn. People went out to give the soldiers flowers and food. Teenage boys scampered after the soldiers, cheering and laughing. Paris was in a festive mood, because the soldiers brought with their song a message of hope.

But the political situation is very contentious. Papa read a copy of the Brunswick Manifesto, a declaration written by the Duke of Brunswick, the commanding general of the Austro-Prussian Army. The duke warned the citizens of Paris to obey the king. The declaration is menacing; it threatens with violent punishment if people do not obey.

The Brunswick manifesto responds to the one from the revolutionary groups who oppose Louis XVI, those who do not recognize him as sovereign of France. The Assembly is offended by the duke's declaration and ordered the sections of Paris to get ready for war. Yesterday, Maximilien Robespierre called for the removal of King Louis. Robespierre is a deputy to the Convention, a lawyer and leader of the most radical political group against the monarchy. After his speech, some Paris sections demanded dethronement.

Father is in his library with some colleagues, speculating what will happen next. The manifesto created both fear and anger among Parisians. The revolutionary leaders are inciting the people of France to revolt. I am afraid of the consequences.

Sunday—August 5, 1792

Violence escalates. Millie came in this morning very agitated and hysterical, almost in tears, with disturbing news. A group of sans-culottes were ransacking the home of M. Comte Voyeaud and there was a big commotion in the streets. Millie described the looting and the pandemonium she witnessed. Mother had to give her a cup of tea to calm her down.

A large group composed of men and women ransacked the château, taking everything they could. Millie ran as fast as she could to tell us. The mob threw out the family's clothes, furniture, and paintings. Many of the lawless men were drunk, shouting madly, inviting people passing by to get the stuff. We wondered about the family of M. Voyeaud's family. And since they live just a few blocks away, I feared that the ruffians would expand their pillage to our street. Millie ran downstairs to secure the tricolor ribbons on all doors and windows.

Now the city is quiet. The bells of Notre Dame just sounded the midnight hour. What plot is being conceived in the minds of the revolutionaries right now? Only time will tell.

Friday—August 10, 1792

The sounds of the alarm bells shattered the stillness of the summer night. I was in bed when the peal of the church bells roused me. I dashed to the open window, peering into the night while the bells sounded the call to arms. The tocsin was ringing and ringing, and soon the angry people of Paris would answer, marching against King Louis.

I rushed to join my parents when I heard voices and the footsteps of people gathering in the street below. Torches illuminated the stagger of those who answered the call though the dark streets. The points of light seemed like a swarm of fireflies buzzing with rage, moving towards the Tuileries Palace where the king slept unaware of the danger. I shivered at the sounds of enraged voices trailing away. I had a sense of foreboding in my chest, but I could not imagine what was to happen later.

My mother kept muttering, "*Mon Dieu! Mon Dieu!*" and as firmly as she could, sent us back to bed. I could not sleep, not even after the church bells stopped ringing and the voices in the distance became silent. Beneath the relative quietness, fear enveloped the city. I was up long before the first light of dawn glimmered on the horizon, and I saw from my window a dark red sky, a view of the morning I had never seen before. It was a premonition of the blood that would flow hours later.

After breakfast, while helping Maman fold the linen, the sudden firing of firearms in the distance startled us. The clatter was terrifying and I felt as if lightning hit me. My sister Angelique ran into the arms of our mother. The angry mob on Rue des Antoine was moving on their way to attack the Tuileries Palace.

It was terribly hot, but Mother kept the windows closed and the curtains drawn all day; the sounds from shooting and screams were not muffled by the still air. My father came early, saying the revolutionaries had taken the royal château. King Louis and his family fled to the Legislative Assembly building before the furious mob arrived at the Tuileries. A violent confrontation broke out across the gardens, with groups of citizens fighting the king's Swiss guards and the National Guards.

In the attack, the Swiss guards, the maids, and cooks of the palace were butchered. Papa described how the Champs Élysées and the Tuileries gardens were covered with corpses and blood. We were trembling just imagining the horror. Maman was sickened, repeating with shaky voice, "*Mon Dieu! Mon Dieu!*" The attack on the palace was brutal, as if all the hate and revenge boiled over, destroying any sense of decency in people. The royal family was unharmed in the assault. However, they are now under house arrest.

After dinner I went to my father's library. But I could not concentrate, feeling the tension of the city that lingered long after the attack. I know the killings continued into the night and I can't help it but get tormented by just thinking about it. The fire that started earlier in the Tuileries Palace is still burning and the smoke seeps through the closed windows. How long can we bear this madness?

France is crushing its king. Today the leaders of the revolution publicly condemned King Louis XVI and Queen Marie Antoinette. The royal family is imprisoned in the tower of the Temple, including the children and the king's sister. The sans-culottes got their wish: to humiliate, defeat, and send Louis XVI to prison.

What has he done so horrible that the deposed king is thrown into a dark dungeon, where only the worst criminals belong? It's an injustice. I feel sorry for the royal family; even if Louis XVI was not an efficient ruler, this is no way to seek justice. Long ago, Queen Marie Antoinette alienated the affections of the French citizens. Rumors circulated for years that she used to host lavish fêtes while the peasants were starving just outside the gates of Versailles. Her enemies claimed that Marie Antoinette went through gowns and jewels as if they were limitless resources and rights of her passage.

Even if true, those excesses do not justify the punishment. I am sorry for the family, how can anybody not pity them? Princess Marie Therese is only fourteen years old, just like my baby sister, and the charming Dauphin is only six. Children should not be thrown into a prison. What is their crime? Doesn't anybody in the government feel any compassion for them?

I wish there was a way to stop the violence that sweeps my beloved Paris. Arrests of innocent people have become commonplace. A few days ago, the Abeé Sicard, a teacher and protector of deaf-mute children, was imprisoned in the Abbaye along with many other priests. Students and teachers from the school have gone to the Assembly to plead for his freedom. What has he done to be arrested? Nothing. Papa says that his only infraction is being a priest who resists to be ruled by the new government. Is this the revolutionaries' ideal of justice?

The alarm bells awoke us. "The Prussian army is advancing!" yelled the town crier early this morning. The call to arms sounded and the city gates closed in haste. Drums beat all day, people run in the streets like mad, and children wailed amidst the chaos. I saw through my bedroom window soldiers on horseback, trying to keep the peace in a frightened city. The smell of gunpowder sifted through the closed windows along with the hysteria of war. It was a scorching hot day, but the tension of the impending attack on the city was more oppressing than the air. We spent the long hours in a state of disquiet. I tried to distract my mind, finding refuge in my books, to see me through the rigours of the siege.

It is already past midnight and the city is quiet. But what terrible things are happening now, hidden under the darkness?

Saturday—August 25, 1792

Looking at the stars in the dark night, my mind wanders, taking me back to numbers. When I began to study prime numbers, I could not find a good explanation as to why the number 1 was not included in the set of prime numbers. The books simply list the first prime numbers starting with 2, 3, 5, 7, 11 and so forth.

Well, now I know why. There is a very important theorem called the Fundamental Theorem of Arithmetic, which states: "Every whole number greater than 1 can be expressed as a product of prime numbers *in one and only one way.*"

I take any number, and write it as the product of its factors. For example, the number 6 can be written as:

$$6 = 2 \cdot 3$$
also $$6 = 1 \cdot 2 \cdot 3$$

According to the Fundamental Theorem of Arithmetic, only the first expression is valid. This is the only one way to express 6 as the product of prime numbers. That explains why 1 is not considered a prime number.

Is there a way to prove the validity of this theorem for any number without having to check arithmetically one by one? Yes, Euclid gave a proof in one of his books of the famous *Elements*. I must learn how he did it and then attempt my own proof.

Thursday—August 30, 1792

Paris continues under siege. The Commune, the new revolutionary government, is using terror tactics to control the citizens. That includes authorized home searches. If the Commune suspects someone is a conspirator or a royalist, they send a group of soldiers to the home to arrest him or her. They call the searches "domiciliary visits," as if they were friendly calls, but in reality the sans-culottes hope to uncover traitors and conspirators who could threaten the nation

Anybody could be targeted for a home search, even innocent people like us. The leaders of the Commune justify this intrusion as a necessary measure to seek out hidden firearms and apprehend suspects; they look for incriminating evidence to use against a citizen. The soldiers can show up late at night or in the early morning to ensure everybody is at home. The search party includes ten or more heavily armed men with sabers, pikes, and guns, and they rummage around a house, not leaving anything unturned. Most arrests so far are in the district Montagne Sainte-Genevieve, but I wonder how long before we see the search parties here. How do we know whose home will be next?

Sometimes I hear Papa and his friends carrying on their political debates, and they warn each other to keep quiet in public. One citizen can denounce another, even with little or no evidence. How many friendships have crumbled because of treachery? Everybody is afraid of everybody. I hope Papa has good loyal friends who share his ideology, and who will not betray him.

Father brought me a new book that Monsieur Baillargeon was saving for me in his bookshop. But the text is written in Latin and it will be a struggle to translate it. So far I translated the title: *Philosophiae naturalis principia mathematica,* as "The Mathematical Principles of Natural Philosophy." M. Baillargeon told me that the gentlemen who purchased this book referred to it as *Principia.* The author is Sir Isaac Newton, considered one of the greatest mathematicians of the last century. I find the subject difficult and the text hard to understand. The *Principia* contains so many concepts and ideas so new to me.

It's hard to remain calm in these uncertain times. Right now I feel a bit apprehensive about staying up late, afraid that the Commune might send a search party to my home tonight. What would I do if heavily armed soldiers would suddenly appear in my bedroom? For sure Papa would not allow them to harm me, but what if he is arrested? Oh no, I do not even want to think about it. Perhaps it is best that I do my studies during the day.

Monday—September 3, 1792

A call to arms led to more massacres! About one o'clock yesterday afternoon the tocsin startled us. It was followed by gunfire and the beating of drums. A member of the Commune on horseback raced through the streets, proclaiming the enemy was at the city gates. The Prussian army had captured Verdun. We saw through the windows people running, assembling in different parts of the city, armed with pikes, pistols, and a gruesome anger showing on their faces.

Papa later described to Maman the horror that ensued after the alarm sounded. At about seven in the evening, enraged mobs surrounded the *Église des Carmen* and massacred all the priests incarcerated there. Then the wild crowd assaulted the prison of the Abbaye and killed more people.

Mother was crying upon hearing such barbarity. She kept muttering, *Mon Dieu! Dieu le Père!* Even Papa, usually so calm and controlled, had to take a moment before concluding that he had no

186

words to describe the shocking cruelties, the bloody spectacle, the madness of the people he was unfortunate to witness. "But why kill the priests?" Mother inquired. We knew that there was no acceptable answer to justify the slaughter. The sans-culottes believed that the prisoners would escape to help release the imprisoned King Louis and start a counterrevolution. Their anger and distrust of the monarchy has clouded their reason.

This morning, as Madame Morrell was assisting Mother with her coiffure, Millie came in, breathless and shaking with fear, to tell us that the dead bodies from the church massacres were laid on the Pont Neuf to be claimed. She had to walk by the bridge on her way to the market and witnessed the macabre spectacle. I closed my bedroom window, afraid to get a glimpse of what could only be a most gruesome nightmare unfolding so close to home. Mother immediately cancelled her visit to Madeleine and instructed Monsieur Morrell to drive Papa using another route away from the bridge. But it was too late; the coachman replied that they had already seen the horror Millie was describing.

The pealing of bells continues, echoing through the air, day and night, like desperate cries for help to the heavens. Is anyone there listening?

Saturday—September 8, 1792

Madame de Maillard came to visit this evening. When she arrived I was reading, struggling to translate a page in Newton's *Principia*. After she kissed my cheeks, Madame addressed me affectionately as "Sophie, the mathematician." It is the best compliment anybody could ever pay me.

Glancing at the book before me, Madame sat to tell me the same book was translated into French, by a remarkable woman! She told me that over forty years ago Gabrielle Emilie le Tonnelier de Breteuil, Marquise du Châtelet, translated Newton's book. The lady scholar also wrote new sections that explain additions and corrections that were made later to the *Principia*. With her notes, the Marquise clarified many difficult concepts in the original book.

187

Before Mme de Maillard finished telling me the good news, I was already planning to search for the translation of Newton's book. Madame added a bit more about the Marquise du Châtelet. Here I summarize what she said.

Gabrielle Emilie le Tonnelier de Breteuil was born in 1706 in Paris. Her father, the Baron de Breteuil, was principal secretary and introducer of ambassadors to Louis XIV. As a child, Gabrielle Emilie had tutors, including her father who instructed her in Latin. Mlle de Breteuil was very intelligent and had a high aptitude for languages, mathematics, and the sciences. By the time she was twelve years old she could read, write, and speak fluent German, Latin, and Greek. At age nineteen she married the Marquis du Châtelet and had three children.

Later, Madame du Châtelet hired tutors to teach her geometry, algebra, calculus, and sciences. She was interested in the work of Isaac Newton, so she asked Moreau de Maupertuis, a member of the Academy of Sciences, to teach her the theories of Newton. An eminent mathematician himself, Maupertuis supported Newton's ideas that were, at the time, debated by French scholars.

Mme du Châtelet wanted to attend the regular Wednesday meetings of the Academy of Sciences in the king's library at the Louvre where the latest scientific knowledge was discussed. However, women were not allowed to attend these meetings. She also wanted to meet with colleagues for discussions at a coffeehouse that was popular with scientists, philosophers, and mathematicians.

But women were also banned from coffeehouses and so she was denied entry at Gradot's café. Since the reason given was that no person wearing skirts were allowed, Mme du Châtelet had a suit of men's clothes made and entered the coffeehouse to meet Maupertuis and other scholars. They ordered a cup of coffee for her, treating her as just another member of the group. The proprietors pretended not to notice that they were serving a woman, and from then on Mme du Châtelet could walk in and nobody bothered her.

When she was twenty-eight years old, Mme du Châtelet met the philosopher Voltaire. They moved to her Château de Cirey to write about science and philosophy. Their collaboration was as close as their friendship. In the *Introduction to the Elements of the Philosophy of Newton*, Voltaire stated that he and Mme du Châtelet worked together in the writing of this book. After this joint project,

Mme du Châtelet continued her study of mathematics and completed her translation of Newton's *Principia*. Alas, she died before her translation was published.

It is a fascinating story. I am so glad I had a chance to learn about an incredible woman scholar. But now it is important that I find Châtelet's translated version of Newton's *Principia*. Monsieur Baillargeon must not have a copy, or he would have told me already. I will ask Papa to help me locate it in another bookstore.

Monday—September 10, 1792

Violence and strife erupt in the streets of Paris everyday. We live in terror. The gates to the city are closed, armed soldiers patrol the streets, houses are searched for weapons, and the fear of persecution is felt everywhere. Maman has placed tricolor ribbons across doors and windows "to keep away the bloodthirsty crowds roaming the streets," as she says. The tricolor ribbons are a symbol of our loyalty to the republic, hoping such display will keep us safe from the extremist revolutionaries. We spend most of the time on the upper floors, fearing for my father when he goes out to tend his business. Millie's errands are limited to the essential trip to the bakery and the market.

More attacks on the prisons continued after the first massacre of prisoners. It is not just the priests in the prisons who have fallen victim to the angry mobs. Many other citizens are murdered as well. At the Bicetre prison many of the people killed were younger than eighteen years of age! Father says that most of the killings are preceded by a trial, run by a mob court devoid of any justice. The so-called judges in these courts are the actual killers themselves, and most of the time they are drunk. The sight of the killers described is revolting—covered in blood and laughing.

The most hideous killing was that of Princesse de Lamballe, the dearest friend of Marie Antoinette and former superintendent of the queen's household. She was imprisoned just for being close to the queen. During her trial the young lady refused to denounce King Louis and Queen Marie Antoinette. This enraged her accusers. Then the princess was stripped, raped, and her body was mutilated. Why? What could possibly cause such a horrible monstrosity?

My father blames the massacre on the revolutionaries' fear of a counterrevolution. They said they have to defend Paris by watching for rebels and traitors that oppose the revolution. But why the massacres? Why such inhumane brutality? The Princess de Lamballe was innocent, and yet they killed her like an animal, and for what? What is the murderers' justification?

<br>

Saturday—September 15, 1792

I am inspired by the Marquise du Châtelet. I envy her strength and bold spirit. She was privileged to have had tutors and opportunities to pursue her intellectual ambitions. But she also had to cope with the prejudice of a society that did not take women scholars seriously. I am impressed by the range of her accomplishments. She translated into French several classical works written in Latin and other foreign languages. For example, she translated *Oedipus Rex*, a play by Sophocles written in Greek.

Madame du Châtelet also translated *Institutions de physique*, which was an explanation of the metaphysical theories of the German mathematician Gottfried Wilheim von Liebniz as expressed in his *Monadologie*. And of course Madame du Châtelet's greatest contribution to France is her translation of the *Principia* by Isaac Newton from the original Latin into French. I can't wait to read it.

<br>

Thursday—September 20, 1792

I have the *Principia* in French. Papa found for me the translated version of Newton's book: *The Mathematical Principles of Natural Philosophy*. I read a full chapter, but I still cannot understand much of it. So far I can only summarize some of Newton's ideas. The *Principia* begins by defining the concepts of mass, motion, and three types of forces: inertial, impressed, and centripetal. Newton also gives definitions of absolute time, space, and motion, offering evidence for the existence of absolute space and motion.

Newton proposes "three laws of motion," with consequences derived from them. The remainder of the *Principia* continues in the form of propositions, lemmas, corollaries, and scholia that are quite difficult to comprehend. Book One, *Of the Motion of Bodies*, applies the laws of motion to the behaviour of bodies in various orbits. Book Two continues with the motion of resisted bodies in fluids, and with the behaviour of fluids themselves. In Book Three, *The System of the World*, Newton applies the "law of universal gravitation" to the motion of planets, moons, and comets within the solar system. He explains a diversity of phenomena from this unifying concept, including the behaviour of the tides, the precession of the equinoxes, and the irregularities in the moon's orbit.

Even the translated version of the *Principia* is a very difficult book indeed, but I've made a commitment to study it. I intend to learn Newton's science one day.


Sunday—September 30, 1792


I am depressed. Although I try not to think about the massacres of the past several weeks, I can't help it to worry about Papa. He and his colleagues talk about the increase of violence and the arrest of innocent people charged as counterrevolutionaries. They say the prisons are full of falsely accused citizens. I try not to listen, but it is difficult to remain oblivious to the social and political horrors happening every day.

To brighten my mood Papa brought me a new book. It's a calculus text entitled *Leçons sur le calcul differential et le calcul intègral* (*Lessons on Differential Calculus and Integral Calculus*), authored by Monsieur J. Cousin, an acquaintance of my father. Although I am making good progress, this book may help me understand Newton. There is so much I need to learn.

191

Calculus analysis deals with the rate of change of quantities, which I interpret as slopes of curves, and the length, area, and volume of objects. This book divides the calculus into differential and integral calculus. The first deals with derivatives of functions, and the second with integrals of functions. I have to study calculus as earnestly and diligently as I have studied other topics. Before I begin, it would be interesting to learn what led mathematicians to develop this branch of mathematics.

Saturday—October 6, 1792

Papa brought me another book on calculus written by Maria Gaëtana Agnesi, titled *Instituzioni analitiche ad uso della gioventù italiana*. He told me that he bought it because the Marquis de Condorcet recommended it for me. The book has many examples to illustrate the methods of calculus. The scholars at the Académie des Sciences have praised it as the best instruction book in differential calculus.

Maria Gaëtana Agnesi is an Italian mathematician. In the preface of her book, Agnesi thanks a person named Rampinelli for helping her study mathematics. She wrote: *"With all the study, sustained by the strongest inclination towards mathematics, that I forced myself to devote to it on my own, I should have become altogether tangled in the great labyrinth of insuperable difficulty, had not [Rampinelli's] secure guidance and wise direction led me forth from it ...; to him I owe deeply all advances (whatever they might be) that my small talent has sufficed to make."*

She was so fortunate to have had Rampinelli to teach and guide her studies.

Saturday—October 13, 1792

My father met with his friends this evening. They discussed the latest political events and the effects of the revolution on the lives of Parisians. As I spoke to Madame Geoffroy, several voices at unison urged me to address her as *"Citoyenne Geoffroy"* (Citizeness Geoffroy). At first I was perplexed and was not sure I understood correctly, but Papa then explained that from now on the titles monsieur and madame will be replaced by the appellations *citoyen* (citizen) and *citoyenne* (citizeness). According to him, it was officially decided by a resolution issued by the Paris Commune last week. Monsieur LeBlanc resumed the discussion, adding that the titles madame, mademoiselle, and monsieur are no longer acceptable. The new forms citizen and citizeness are intended to erase the distinctions between the various classes of society.

It makes no sense. The new form of address is insufficient when it comes to women. If a person says, for example, "Citizeness Germain," who would know whether they are talking to my mother, a married woman, or addressing me? I do not dispute the use of "citoyen" to address males, after all, the social title "monsieur" is applied to all men, but I disagree with the use of "citoyenne," for it does not apply to all women. So, as far as I am concerned, my mother always will be "Madame Germain," and I always will be "Mademoiselle Germain."

I must ask, furthermore, how can anybody presume that abolishing courtesy titles makes us all equal? The new appellation rules will not change how people treat one another. There will always be prejudice amongst people, whether we address each other with title or without.


Monday—October 15, 1792


Paris continues under siege. Violence, terror, and scarcity of goods punctuate our lives. Tallow is in short supply. We are only allowed a few candles per week. Now I have to study by daylight. Sometimes I do not realize how dark it is, until the shadows of the night make the letters in my books disappear and my eyes strain to see. Millie and Angelique offered to give me their own candles, but it would be selfish on my part to deprive them of light, especially now that the nights are getting longer.

I am teaching myself differential and integral calculus. It is quite a challenge. Calculus, like other parts of mathematics, was developed by the need to explain physical theories. My research suggests that two scholars in different parts of the world came up with similar ideas simultaneously. But the way each scholar arrived at their calculus is so different.

It is assumed that Isaac Newton developed his ideas of calculus when the plague closed the schools in the summer of 1665. Newton returned to his home in Lincolnshire and, in seclusion, he began to lay the foundations of the new mathematics. The "method of fluxions," as he termed his calculus, was connected with the study of infinite series. A "fluxion," expressed by a dot placed over a letter, such as $\dot{x}$, was a finite value, a velocity such as $\dot{x} = \dfrac{dx}{dt}$, where $x$ represents distance and $t$ represents time. The letters without the dot represented "fluents."

The method of fluxions was developed on Newton's insight that the integration of a function is merely the inverse procedure to differentiating it. Taking differentiation as the basic operation, Newton produced simple analytical methods that unified many separate techniques previously developed to solve apparently unrelated problems, such as finding areas, tangents, the lengths of curves, and the maxima and minima of functions. Because his new mathematical ideas were so different, Newton feared criticism so he did not publish his memoir until 1704. It appeared as an appendix to his book on *Optiks*. Newton claimed to have written *De methodis serierum et fluxionum* in 1671, but for some unexplained reason he did not publish it until much later.

The German mathematician Gottfried Wilhelm von Leibniz had similar ideas. Leibniz corresponded with French mathematicians, and one day he came to Paris to work at the Academy of Sciences. It was during this period in France that Leibniz developed the basic features of his version of calculus. In 1673, he was still trying to develop a good notation for his calculus. A few years later he wrote a manuscript using the integral $\int f(x)\,dx$

notation for the first time. In the same manuscript he gave the product rule for differentiation.

By 1676, Leibniz discovered that $d(x^n) = nx^{n-1} dx$ for both integral and fractional $n$. For some unknown reason Leibniz did not publish his new mathematics for a long time. In 1684, he published his differential calculus in a paper entitled *Nova methodus pro maximis et minimis, itemque tangentibus* (*A New Method for Maxima and Minima as well as Tangents*), published in *Acta eruditorum*, a journal he founded in Leipzig two years earlier. The paper contained the $dx$, $dy$ notation, the rules of differentiation, including $d(uv) = udv + vdu$, and rules for computing the derivatives of powers, products and quotients.

Mathematicians in France now use Leibniz's notation—derivatives expressed as $dy/dx$— rather than the fluxions developed by Newton. Leibniz was also the one who called this branch of mathematics *calculus differentialis* and *calculus integralis*. In 1696, the first textbook on calculus appeared. The *Analyse des infiniment petits* was written by the French mathematician Marquis de L'Hôpital, who had learned calculus from the great Swiss scholar and mathematician Johann Bernoulli.

Perhaps one reason why I have had difficulty understanding Newton's book is due to notation. I will ask Papa to help me find Leibniz publications. L'Hôpital's textbook may be even better since it is written in my own language. I must send message to M. Baillergon to locate this book for me.

Friday—November 2, 1792

I wonder about the future of my beloved France. The brutal killings and the social chaos have changed my country. Right now there are two competing views on which direction France should go, embodied by two political parties: the moderate Girondists who favor a peaceful reconstruction of France, and the more radical Jacobins, led by Maximilien Robespierre, who favors purging France of its imperial past. Who is right?

Meanwhile at home we try to maintain a sense of normality. We go about our lives as if nothing has happened, but I know this is only superficial. We all have witnessed the horrors brought about by the misery of the poor, and the misguided eloquence of the political leaders. Even my little sister has seen death in the face; she no longer runs to Mother when the mobs are heard outside our windows, searching for a victim, searching for justice and a truth that seems to elude them.

We all have grown beyond our years. Mother's beautiful face is lined, and a mask of worry and apprehension seems to hang over her eyes. Her smile is not as easy as it was years ago. My parents no longer go to the theater, and we visit only our closest friends. But we try to maintain our family traditions. We have our literary evenings and read plays by Molière and passages from Rabelais. Every week Angelique regales us with a beautiful piano concert, and Mother sings her beloved arias. Millie joins us at these gatherings. She now reads so well and volunteers to recite popular poems.

I have my books to see me through the rigours of a siege that doesn't end. My books, they are quietly waiting to reveal the knowledge imprinted in their pages, opening a door to a world without violence.

Thursday—November 15, 1792

King Louis is ill. The newspaper reported that his health is quickly deteriorating after living imprisoned in that dreadful cold prison. Our morning prayers were dedicated to him. My mother recited the seven psalms for his recovery. I am still confused, trying to understand why the king and his family are prisoners in the Temple. What is their crime? What about the children?

I am glad that my father is no longer a deputy in the government. Papa resigned a long ago, even before the royal family was locked up, because he did not want to share the blame for crimes and follies, which he does not condone. After so much injustice and violence, Papa had no option but to leave the government.

I've struggled with the concept of differential $dx$. If $\Delta x$ represents a distance along the $x$-axis, or the difference between any two values of $x$, $dx$ means exactly the same thing, with one difference: $dx$ is a differential distance, a very, very tiny difference. If I have a quantity whose value is virtually zero, then $2+dx$ is pretty much, well, 2. Or if I divide a number by $dx$, such as $3/dx$, the result is infinity!

There are two cases under which terms involving $dx$ can yield a finite number. One case occurs when I divide two differentials. If they are the same, we get the quotient equal to 1, such as $dx/dx = 1$. However, if the differentials are different, the quotient can be some finite number since the top and the bottom are both close to zero, such as $dy/dx$ = constant, or any value. The other case is when I add up an infinite number of differentials, the sum is equal to some number greater than zero and less than infinity. It makes sense. These two cases describe the derivative and the integral, respectively.

*Eh bien.* The flame of my candle flickers nervously, casting shadows on the walls like ghosts and apparitions. I must bid good night before the flame expires and leaves the room in total darkness.

The health of the king is worsening. And yet, he remains imprisoned in the Temple, deprived of all bodily comforts. What is going to happen to him and his family? Father and his friends consider whether Louis XVI can be tried in court. Many of the revolutionary leaders have accused him of crimes against the nation. However, the Constitution of 1791 protects the monarch from any penalty worse than dethronement, and no court in the land has legitimate jurisdiction over the king.

Father is not sure there are enough citizens in Paris to use this argument anymore. All his friends insist that King Louis cannot be condemned without a trial. Of course, Papa is on the side of the monarch, but he feels that, with no other alternative before them, the Convention may just take on the role of a court. I do not understand; this is contrary to accepted judicial principles. I do not comprehend the law, but surely this does not look good. Mother keeps praying, hopeful that somehow this nightmare will end. I am afraid the revolution has gone past the point of no return.

Friday—December 1, 1792

I am furious! Mother called me to her sitting room to give me a lecture on manners. What made me angry is that she told me to be especially courteous with Antoine-August, who will come for dinner with his parents next week. Maman reminded me to behave like a nice young lady. For Maman, being a lady translates into conducting oneself like a naïve, helpless woman and allow a man to feel superior. I can't do that. Why are women expected to be the weaker sex and be subservient to men?

I cannot stand to talk with Antoine-August anymore. He has become conceited, condescending, patronizing, and acts as if he knows everything. He intersperses mathematical terms in his speech to appear as if he knows a lot. But he doesn't fool me. Once I tried to correct him when he stated something so absurd that even I who am not a trained mathematician knew better, but he acted as if I was the stupid one, and he stopped me before I could make him aware of his error. I was so angry that I didn't have the words to debate him. Antoine-August can impress people with his airs of intellectual prowess and his sophisticated words, but I can't stand his arrogance! Sometimes he speaks nonsense; once, he even confused irrational numbers with fractions! But he will never admit it.

After Mother's lecture I was even more rattled because she also selected for me the brown dress to wear at the party. I dislike that dress! I was sulking in my room when Millie came in, and to console me she sang a song that says: "*Il faut bien que je supporte deux ou trois chenilles si je veux connaître les papillons*" (I must put up with two or three caterpillars if I want to know butterflies). After a few minutes, I could not help but smile at Millie's silly song. Perhaps she is trying to tell me something. Now that I think about it, maybe it would benefit me to be friends with Antoine. He could tell me about the topics he is studying at school, especially if he thinks I would be impressed. Maybe he would let me see his lecture notes in mathematics.

I will be polite to Antoine-August as long as he does not treat me condescendingly. I set a limit on that.

Friday—December 7, 1792

Why women do not demand an education like men do? Boys have tutors to teach them science and mathematics, and then they go to the schools of high learning to pursue degrees. As scholars men correspond with others in their fields of research, and learn from one another. Men can become lawyers, doctors, mathematicians, or anything they want to be. However, women are excluded from these professions; they stay home and raise children. Even if we are smart and wish to learn sciences and mathematics, we are told it is a right reserved to men only; we are expected to be content knowing only the fine arts and literature, but some areas of knowledge are "not suitable for women." Why?

When I solve an equation or prove a theorem, I feel a sense of accomplishment, an intellectual pleasure, but I am still Sophie, a girl. If I can think mathematics, then it means mathematics is not reserved for the minds of men only. Who decided to exclude women from this pursuit? Women have demonstrated that they are as intelligent as men. Many women in history have made important contributions to the scientific development, starting with Hypatia of Alexandria in ancient Greece. I know of two other women scholars in this century, the Marquise du Châtelet and Maria Gaëtana Agnesi. There must be many more women as intelligent as these women, and I am sure now there are many more like me who aspire to scientific learning, but we are repressed by society and excluded from schools.

Oh, I would give anything to have the privilege to study with a mathematician. I have learned so much in the past three years, but not enough! Because, even though I have struggled to feed the hunger of my intellect, I have become fully aware of my status as a woman in our society. The servitude and prejudice under which society keeps women oppressed is humiliating. I hope that the new Republic of France will justify itself by offering women opportunities for education equal to men's. The social demands of the people were heard but, as of today, education is still not a right of women.

Friday—December 14, 1792

I am anxiously waiting for a letter from Antoine-August LeBlanc. He and his parents came to dinner a week ago, and he promised to send me some of his class notes. I must say, I did not expect this turn of events. At the party, as always, the conversation began with politics and the state of affairs in Paris. It continued with discussions about the royal family imprisoned in the Temple, and the impending trial of the king. Antoine, at seventeen, is now an adult, so he spent the evening arguing with the gentlemen about the case, expressing his fervor for the republic.

I waited for the opportunity to talk with him because I had so many questions to ask about his studies. It wasn't until midnight, when they were about to leave, that I gathered the courage to approach him. He was surprisingly attentive to my questions, quite courteous, and he even offered to send me copies of his lecture notes. He mentioned that he is studying differential equations. He claimed that by knowing calculus one could understand the world. Well, I wish to know it too. And I shall learn calculus as well.

The following day I went to the bookshop and asked Monsieur Baillargeon to find me books on differential equations. This afternoon M. Baillargeon had for me several books authored by Leonhard Euler. The books were not too expensive because they're used, but they are in excellent condition. I was delighted and without a second thought I bought them all. One of the books is titled *Institutiones calculi differentialis*. The other is a three-volume book,

*Institutiones calculi integralis.* Now I have the best books to learn differential and integral calculus.

Euler included a theory of differential equations, Taylor's theorem with many applications, and many more concepts I am eager to learn. I have just read the section on differential equations where Euler distinguished between "linear," "exact," and "homogeneous" differential equations. I spent a day reviewing functions, which Euler presented in a clearer manner.

Now I am beginning to understand this branch of mathematics. Calculus is useful for answering questions in many sciences. The theory of differential equations has an inherent beauty because it provides us with a unique tool for understanding nature. There is so much more I need to learn, but I am determined to master calculus by the end of next spring.

Tonight I look up at the sky; there is no moon in sight. It's past midnight already. The only sound I hear is the howl of the wind outside my bedroom window. I can't help it but to think about the king, the queen and their children, locked up in that cold dark prison. What future awaits them?

Thursday—December 20, 1792

Paris has changed so much. When I was a little girl I looked forward to this time of year full of joyful anticipation. The white snow, the open fires, the preparation for Christmas; the prelude celebrations used to transform Paris into a city of light. We used to go to church, where a thousand candles were lit. I enjoyed so much the pastoral festivals that announced the approaching celebration of the Nativity of Jesus. Even the poorest citizens would look at the holidays with a sense of hope because the wealthy ones were more generous, giving away presents to those who had none. Those blissful days are gone.

The cold weather is here again, but this time no fire can warm the hearts of the people. Something changed since the day the monarch of France was arrested as a common criminal. Now his trial is unfolding as a mockery to justice. My mother prays for a miracle, but the king's lawyers cannot defend him against the terrible accusations.

The trial is a parody. Many people have already condemned him. Why doesn't anyone shout that this trial is illegal? King Louis is charged with treason. If the jury finds him guilty, the legal penalty is death. Some of Father's colleagues argue that to allow the king to live would undermine the principle of revolutionary justice. What justice? Murder is committed in the name of justice. No, there is no justification for the bloodshed that now soils the ground of my beloved Paris.

I fear for my father. Although he stands for change and supports the revolution, he is sympathetic of the king. If he could vote, he would not condemn Louis XVI to die. I am sure of that. My father is outspoken, not afraid to stand for what is right. But there are many hypocrites who betray their friends, even their own brothers. My mother fears that one day somebody may accuse my father. But, as he says, there is no sense in worrying about what has not happened. And he goes on fearlessly speaking against injustice.

Tuesday—December 25, 1792

I had a wonderful dream last night. When I woke up I felt so happy, with the remains of the vision still lingering in my mind. It was a vivid dream, like a revelation. I was floating, flying far away from the Earth, where the darkness of the sky was only illuminated by the millions of stars shimmering in the distance. My body felt like a feather, weightless and free. As I soared, I spotted mathematical equations and symbols floating in all directions towards and away from me, appearing and disintegrating like soap bubbles.

There were letters $x$, $y$, $z$, and numbers $\pi$, $e$, $i$, 1, dancing like butterflies in a flower garden. When I'd reach out to take one equation, it would fly away farther from me and then another would appear near by. There were many kinds of equations, some simple ones I recognized, but other equations were complicated and I did not understand them. There was one particular equation that I needed to have, so I chased it with all my might. When I held the equation in my hand, I felt much joy, a happiness that went beyond laughter. The equation was not a real object, but it felt like a deep indescribable emotion; although I did not touch it, I sensed it, making my heart beat fast.

Upon awakening I could still feel the euphoria. My hand was still holding the book I was reading last night. I got up quickly, trying to remember the equation because I wanted to write it down to try to understand its significance, but I could not recall it. The vision of the equation was gone. What remained in my mind was just the feeling of sheer pleasure I felt holding the exquisite equation.

I am thinking now, if I could conceive an equation while dreaming, it must reside somewhere in my head. Thus, one day the cherished equation will reappear, not in a dream but in my awakened mind.

CʒƉᴖ

The storming of the Tuileries Palace, August 10, 1792.

Attack on the Palais-Royal—Insurrection of August 1792.

The royal family taken prisoners to the Temple, August 13, 1792.

The massacres of the prisoners, September 2–5, 1792.

*Paris, France*
*Year 1793*

# 5

# Upon the Threshold

A NEW YEAR BEGINS, a sickly child born to a mother cruelly beaten, crushed. The mother is France in 1792, a year that began in strife and ended in sorrow. A year ago there was suffering, people rioting for food, dying of hunger and despair. Then the country was besieged by war.

The summer was no better; the citizens of Paris went crazy and stormed the Tuileries Palace, taking the king and the royal family captive and thrown in a cold, dark prison. The people spat on the faces of the monarchs and their children, treating them all like criminals. Violence, rioting, and more massacres followed. In September bloodthirsty mobs murdered hundreds of innocent people, including priests imprisoned solely because they did not swear allegiance to a civil constitution. Overnight the face of Paris changed. A radical Commune took over the government, and the gates to the city closed.

The trial of the King of France filled our hearts with trepidation. Louis XVI was condemned like a common felon, judged in a courtroom full of disloyal subjects, people who turned against a monarch that was once considered the father of the nation. Today Louis XVI was declared guilty. His lawyers could not convince the jury that King Louis meant no harm to the nation. His accusers will ask for the death penalty.

A new year sneaked up on us without fanfare, as subtle as time when day becomes night; I hardly noticed it. This time there was no joyful celebrations, no gifts, no feasts, no laughing. We await the outcome of the sentence, optimistic that the life of the king will be spared.

The wait is over. For weeks Mother prayed for King Louis, hopeful that he'd be found innocent or given a lenient sentence. We kept anxiously waiting for news, awakening each day wondering when the nightmare of France would end.

Today our beloved King of France Louis XVI was condemned to die. The decent people of Paris are at a loss for words. The radical revolutionaries are the only ones that rejoice. We are stunned. Mother wept when Father told us. Even Angelique stayed in her room drawing, trying to immortalize the present with her charcoal sketches. We all fear an uncertain future.

Father and his friends reviewed the conclusion of the king's trial. The voting process took several days to complete; only a few delegates abstained from voting. At the first vote, 693 deputies unanimously voted the king guilty, and the call for a referendum was rejected by 424 to 283. The death penalty carried 387 to 334 votes. Seventy-two deputies asked for a reprieve, but an extra vote saw their demand rejected by 380 to 310. On each occasion deputies answered individually to their names. That is how we know that Monsieur Condorcet voted against the execution of Louis XVI. Why weren't more people like him?

Mother continues reciting her prayers for a miracle that could save King Louis. Papa says it is already too late for prayers.

My heart feels oppressed, my spirit trembles. I must find refuge in my studies. Mathematics is a world without violence where I can thrive and flourish, untouched by the grim specter of murder and sorrow.

Monday—January 21, 1793

Today is one of the most reprehensible and shameful days for France. Today our King Louis-Auguste XVI was put to death at the guillotine. I must record this day for posterity.

We woke up dispirited knowing that something ghastly was soon to happen. A gray and cold sky greeted us, making me feel more miserable. No one at the breakfast table spoke a word and I didn't feel like eating. Even my sister Angelique was unusually withdrawn. The gates of Paris were closed earlier in the morning; the city was eerily quiet, in an expectant state, as if the crime that was about to be committed was insulated from the rest of the world by a wall of silence.

Later, Papa told us that before his execution, when the drums ceased for a moment, the king addressed the mob: "I die innocent; I pardon my enemies." But soon the beating of the drums drowned his voice and the heavy blade swiftly descended. Papa said with a low voice almost inaudible that the crowds cheered when the head of Louis XVI dropped to the muddy floor. Maman did not want to hear another word about it and left the dining room. I heard her weeping in her chamber, trying to intersperse a prayer between sobs. Maman did not bother to hide her sorrow when my sister and I approached her. The cold rain mimicked the tears that glistened in her eyes. I kissed her cheeks, moist with the salty drops. My sister clung to her, and I left her side to be alone.

Standing in the middle of my bedroom, alone in darkness, I began to cry like a child. I tried not to think, for I did not want to imagine the horror at the scaffold. My tears felt warm on my face; my throat hurt as I tried to muffle a sob. Why? I cannot comprehend the violence and the murder. Why did they have to take the life of a man in that horrendous manner? Why they could not spare him? What kind of justice is this when, after killing the king, people laugh and cheer?

Papa offered a hopeful attempt to comfort us: "A republic founded on the blood of innocent victims cannot last long." My God, I hope he is right and that this nightmare ends soon. My heart feels oppressed and my spirit heavy with sadness.

The life of the city has resumed its course after the execution of our monarch, and many shops opened as usual. But to me, it felt like we had just buried a member of our family. My father stayed home. We just could not suppress our grief, and we mourned in private the murder of King Louis-Auguste XVI.

211

I have pneumonia. My chest hurts when I breathe and my body feels weak. Maman claims I contracted the illness for staying up late at night, studying in the freezing cold. It's possible my bedroom became cold when the fire in my heater was extinguished, and I did not pay attention.

Angelique stays with me most afternoons drawing pictures. She says that I had a fever so high I was delirious, and that I was mumbling strange words nobody could understand. She claims that once I was desperately reaching out for my books, and Papa had to hold me to calm me down. But I do not remember any of that.

The fever has subsided and I am beginning to feel better, just weak from being in bed for so many days. The doctor came this morning and gave me a bitter tonic difficult to swallow. Maman wants me to stay in bed a little longer and has forbidden me to read. Papa asked me to heed Maman's advice by not immersing myself so deeply in my studies that I jeopardize my health. He reminded me that if my body is weak by illness, I would not sustain the strength of my intellect. He kissed my cheeks ever so gently. I've promised him that I will not get up until I feel better.

Millie brought me chicken broth this evening and tried to cheer me up by reciting a love poem she learned, she said, while sitting by my bedside watching over me when I was very sick. She now reads so well, I am very proud of her. Then she told me that a letter addressed to me came in Monday's post from Antoine LeBlanc. My heart began to beat fast, thinking that the letter must contain the lecture notes on mathematics Antoine promised to send me. I asked Maman, but she replied that I must wait until the doctor says I am cured. But I cannot wait. I must get better soon. I've neglected my studies too long.

Wednesday—February 6, 1793

Madame de Maillard came to visit me. She brought me a book titled *Mécanique analytique* written by Joseph-Louis Lagrange,

212

a scholar at the Academy of Sciences. She says that it is a book for *real* mathematicians. I perused it, but I don't know enough mathematics to comprehend this material. All the pages of the book are full of equations. I must admit, I feel quite intimidated by what Lagrange calls "analytic mechanics."

Monsieur Lagrange wrote in the Preface of his book: *"One will not find figures in this work. The methods that I expound require neither constructions, nor geometrical or mechanical arguments, but only algebraic operations, subject to a regular and uniform course."* That's like trying to read a novel written in a foreign language. Just like using words to tell something, the equations tell a lot if one understands their meaning. I'm sure one day I shall learn enough mathematics to understand it. If Mme de Maillard thinks I can do it, then I will.

I've also reviewed the notes that Antoine LeBlanc sent me recently. It is the same material that I found in Agnesi's book on differential calculus. Since I had told him I was interested in prime numbers, he also recommended that I obtain a copy of Adrien-Marie Legendre's memoir *Recherches d'Analyse indéterminée*, published in 1785. He said that this paper contains a number of important results such as the law of quadratic reciprocity for residues, and the results that every arithmetic series with the first term coprime to the common difference contains an infinite number of primes.

This is quite advanced material for a lowly pupil like me, one that was deprived of tuition. Could I ever reach the summit of my aspiration to become a mathematician some day? If I can ever understand the material in these scholarly works, then I will categorically say I am a mathematician.


Monday—February 18, 1793


Paris has changed but my life has changed even more. I was a child when the first signs of the revolution manifested and now I feel so grown. My development is not the chronological effect of time that makes me see my existence differently. It's the social changes all around me. In only a few years, French society was transformed; the old customs and traditions are gone, and with them so many values that we held dear. Even the way we address each other has changed.

King Louis XVI is already forgotten; his memory is now but a shadow that follows the members of the royal family that remain imprisoned. They were put away, out of sight, as if the murderers are afraid to see the specters of their sins. Just like Marie Antoinette, the once beautiful queen of France, the graces of human kindness have faded, aged, discarded and replaced by rough, crude discourse. The monarchy was erased in one stroke, like multiplying by zero.

The decent citizens of France now live in fear. People are afraid of being accused, condemned, imprisoned, and put to death. Many times I pondered what I would do without my studies. I could not bear the harsh reality without the researches in mathematics that take my mind into a world where violence and hate do not exist. Learning provides me with an escape; books soothe my mind and comfort me when anguish oppresses my heart not knowing if my father is in danger. How anxiously we live awaiting his return every evening.

It is terribly cold. The sun shone at midday but not long enough to melt the icy snow on the streets. We now spend more time in the kitchen near the burning stove because there is not enough fuel to replenish all the heaters in the house. Coal is scarce, and lamp oil is terribly expensive. Poverty is rampant in the streets of Paris. Maman and Angelique are knitting scarves for the homeless children, the orphans of the revolution.

Monsieur and Madame LeBlanc came to visit this evening. M. LeBlanc took the conversation away from politics and focused on education. He expressed the dire need to establish a new university to educate scientists and engineers. He indicated that the revolution has depleted France of intellectual power since many engineers and scientists have immigrated to other countries. Besides, he stated, the great schools in France are now closed, and the ravages of war have left the transport infrastructure in need of maintenance and improvement.

M. LeBlanc shared the discussions he had with members of the Academy of Sciences to establish a unique school, where great mathematicians like Lagrange and Monge can train civilian and military engineers. He mentioned other names and details about the

214

proposals for the new polytechnic school, including what kind of courses would be taught in mathematics, physics, and chemistry. Mme LeBlanc said that Antoine should be one of the first pupils at this school. He needs to be trained by the best teachers since he would like to attend the École des Ponts et Chaussées, an excellent engineering school that requires rigorous training in mathematics and science for admission.

Nobody mentioned admitting women to this school, but since the new Constitution of France is supposed to guarantee equality among all citizens, I wonder if someone like M. Condorcet again would raise the issue of women's education. I do hope the new government will make allowances for women to be admitted to the polytechnic school.

Nothing is for sure yet since the Académie des Sciences is closed and there is great confusion about the future of the new republic. I can't help it but wonder if the plan is carried out, perhaps it can open a possibility for me. Wouldn't that be grand if I could enroll? I would do anything to be taught by the greatest mathematicians in France.

I must study hard because, as M. LeBlanc emphasized, the new school will admit only the most gifted students, and they will ensure that by requiring students pass a rigorous entrance examination. I shall prepare myself. Oh yes I will. What else can I do to make my dream come true?

Wednesday—February 27, 1793

I feel strong enough to resume my studies. My mind is clear again to meet the challenges of a new topic that at first seemed insurmountable. I resumed my studies of differential calculus.

There is something magical about *Infiniment Petits*. I went back to the basic definition: "a derivative of a function represents an infinitesimal change in the function with respect to whatever parameters it may have." The simple derivative of a function $f$ with respect to $x$ is denoted either $f'(x)$ or $df/dx$. Newton used fluxions notation $dz/dt = \dot{z}$, but it means the same, so I will use $f'$ or $df/dx$

from now on. Well, I can now take the derivative of certain class of functions because I just follow certain rules:

If my function is of the type $x^n$, I use the fact that $d/dx\ (x^n) = nx^{n-1}$. So, if I have $f(x) = x^5$, its derivative should be $5x^4$. This is easy. If I analyze trigonometric functions $sin\ x$ and $cos\ x$, then I use the definitions $d/dx\ sin\ x = cos\ x$, and $d/dx\ cos\ x = -sin\ x$. At first I did not know why, but I have learned to derive these formulas so that I use them without doubt or question.

Taking derivatives is so easy! I could spend hours deriving more complicated functions. However, I wish to learn also how to see the world through mathematics. I must find the connection between differential equations and physics. I am eager to explore this applied aspect of mathematics.

Let's start with a differential equation, an equation involving an unknown function and its derivatives. It can be relatively easy such as $\dfrac{dP}{dt} = kP$, or a bit more complicated such as the linear differential equation: $(x^2 + 1)\dfrac{dy}{dx} + 3xy = 6x$, or even a nonlinear equation. An equation in which one or more terms have a variable of degree 2 or higher is called nonlinear, for example: $x^5 + x - 1 = 0$. Nonlinear equations are difficult to solve and some are impossible.

First I need to master linear equations. Some mathematicians use the notation $y'$ for the $dy/dx$ derivative, or $y''$ for $d^2y/dx^2$, and so forth. Thus, the previous linear equation would be written as $(x^2 + 1) y' + 3xy = 6x$. I need to keep these differences of notation in mind, since I am studying from five different books.

I studied the properties of differential equations and learned to solve them. Now, I must learn how to apply differential equations. But how do I translate a physical phenomenon into a set of equations to describe it? It is impossible to depict nature in its totality, so one usually strives for a set of equations that describes the physical system approximately and adequately.

In general, scientists first build a set of equations, and then they compare the results generated by the equations with real data collected (by measurement) from the system. If the two sets of data agree or are close, then the set of equations will lead to a good description of the real-world system.

I need a simple example to illustrate this process. Say that I want to predict the growth of population in Paris. To do it, I can use the exponential model, that is, an equation that represents the rate of change of the population that is proportional to the existing population. If $P(t)$ measures the population, I write $\dfrac{dP}{dt} = k\,P$, where the rate $k$ is constant. I observe that if $k > 0$, the equation describes growth, and if $k < 0$, it models decay. The exponential equation is a linear equation which is solved by integration, yielding $P(t) = P_0\,e^{kt}$, where $P_0$ is the initial population, i.e., $P(t = 0) = P_0$.

Mathematically, if $k > 0$, then the population grows and continues to expand to infinity. On the other hand, if $k < 0$, then the population will shrink and tend to 0. Clearly, the first case, $k > 0$, is not realistic. Population growth is eventually limited by some factor, like war or disease. When a population is far from its limits of expansion, it can grow exponentially. However, when nearing its limits, the population size can fluctuate. Well, I think that the equation I use to predict the rate of change of population can be modified to include these factors to obtain a result closer to reality.

Aristotle thought that nature could not be expected to follow precise mathematical rules. But Galileo argued against this point of view. He envisioned the experimental, mathematical analysis of nature to be used to understand it. Newton was inspired by Galileo and later developed the laws of motion and universal gravitation. Newton, Leibniz, Euler, and other great mathematicians then created the mathematics that help us converse with the universe.

Oh, how glorious it is to speak such a language and understand the whispers from the heavens and the world around me.

Thursday—March 7, 1793

Someone said that mathematics is the queen of sciences. This must be since, like art and music, mathematics stimulates the senses and lifts the spirit. My pursuit of mathematics has saved me from fretting and thinking about the social chaos and uncertainty of the future.

My mother began morning prayers with thanksgiving and ended with a plea for resolution. The economy is worsening and food is in short supply. Millie comes home with fewer groceries and no longer complains that the shops are almost empty. In some parts of the city, hungry and desperate people attack butcher shops and bakeries to ensure their meager meals.

Abolishing the monarchy has not brought the changes that were supposed to benefit the citizens of the lower class. The ravages of the social revolution can be seen everywhere. There are still many jobless people, many more hungry, homeless, and illiterate. Poverty is rampant. Isn't this what the revolution was supposed to eradicate?

When darkness descends upon Paris, the false sense of security disappears. Soldiers patrol de streets; the sounds of guns going off mingle with the bells striking the hour. The Commune decree forbids people from walking after ten o'clock at night without an identity card. At night, Paris becomes a dungeon of terror and fear.

Friday—March 15, 1793

I mustered the courage to write a letter to Antoine LeBlanc. I asked whether he would permit me to study his lecture notes on mathematics. Mother would be mortified to learn that I was so bold. I must admit, I also feel a bit anxious, afraid that Antoine will either ignore my letter or, worse, make fun of me and belittle my request. However, there is no turning back. He must have received my letter already.

After my illness, I resumed my study of linear differential equations. Thus far, I've learned that there are many different types of equations with many levels of complexity. Mathematically, differential equations are beautiful and, if they are linear, the methods for solving them are straightforward. For example, a type of

differential equation that I learned to solve is $\dfrac{dy}{dx} + p(x)\,y = q(x)$. It is a first order linear differential equation whose general solution is given   by $y = \dfrac{\int u(x)\,q(x)\,dx + C}{u(x)}$,   where   $u(x) = e^{\int p(x)\,dx}$ is   the integrating factor. I suppose I like these types of equations because they are easy to solve, and the solution involves Euler's number $e$. Differential equations can also be written in the form $y' + p(x)y = q(x)$, where $y' = dy/dx$.

But I discovered that there are complicated differential equations of first and higher order, which are more difficult to solve. There is one particular equation of the form $y' + p(x)\,y = q(x)\,y^n$, which cannot be solved using the approaches I've learned so far. I attempted several methods of solution, but I am not sure my result is correct. That's why I thought of writing Antoine. I enclosed my notes and asked to show my analysis to his tutor.

For now I must stop obsessing over the letter. I wrote it formally and correctly, just as Maman taught me. If Antoine chooses to ignore it, I will forget about it and will find another way to learn advanced calculus. However, if he is friendly, Antoine LeBlanc can be my link to the sources of learning, including the scholars themselves. I must be patient.

Thursday—March 21, 1793

There is so much to learn that at times I feel overwhelmed. But when I solve something challenging, I feel exhilarated and want to continue. Especially when I find equations that relate to a physical problem I understand.

I discovered that a phenomenon as simple as that of cooling objects could be represented by a differential equation. For example, when hot chocolate is poured into a cup, it immediately begins to cool off. The cooling process is rapid at first, and then it levels off. After a period of time, the temperature of the liquid chocolate reaches room temperature. Temperature variations for cooling objects follow a physical law that can be stated as "the rate at which a hot object cools is approximately proportional to the temperature difference between the temperature of the hot object and the

219

temperature of its surroundings." Scientists know this as *Newton's Law of Cooling.*

I can write the cooling law with mathematics. With $T(t)$ representing the temperature of the hot object at time $t$, the rate of change of temperature as $dT/dt$, which is the derivative of the temperature with respect to time, and its relative temperature with respect to the ambient temperature as $(T - S)$, I write:

$$\frac{dT}{dt} = -k(T - S)$$

Here $S$ is the temperature of the surrounding environment, and $k$ is a positive constant of proportionality that depends on the physical characteristics of the object.

The equation is a first order linear differential equation. This type of equation is very easy to solve; it requires that one knows how to integrate. But first, I need a boundary condition, in this case, *an initial condition.* In other words, I need to know the temperature of the object at the beginning of the cooling process, at time $t = 0 = t_0$. I write this initial condition as $T(t = 0) = T_0$.

To solve the differential equation, first I separate the variables and then integrate as follows:

$$\int_{T_0}^{T} \frac{dT}{(T - S)} = -k \int_{0}^{t} dt$$

which gives after integration

$$\left[\ln(T - S)\right]_{T_0}^{T} = -k(t - 0) = -kt$$

or

$$\ln(T - S) - \ln(T_0 - S) = \ln\left(\frac{T - S}{T_0 - S}\right) = -kt$$

This equation can also be written in the form

$$\frac{T(t) - S}{T_0 - S} = e^{-kt}$$

So, the solution to the differential equation I developed is formally given by

$$T(t) = S + (T_0 - S)e^{-kt}$$

*Très bien.* This is very easy! However, I know that differential equations can be much more complicated than the simple linear one I used here. And that is what I must learn next.

Monday—April 1, 1793

Like dusk that separates light and darkness, an instant of time intangible yet real, I am here now, no longer a child, but on the threshold of my womanhood. Today I am seventeen years old. Today I cross a verge into my future. What is my *raison d'être*? What is the reason for my being? What is my purpose in life? I only know that I am eager for genuine knowledge and wish to give myself to the study of mathematics.

I tremble with anticipation at the future before me because, like in my dreams, mathematics will rule my world. I could no longer deprive my soul of the beauty and harmony of the mathematics that intermingles with my thoughts, like the notes in an exquisite symphony. I want to create mathematical ideas and to live in such fascinating realm independent of the material world. I wish to uncover the mysteries that remain beneath unsolved equations. I wish to see nature through the eyes of mathematics.

In the last few years I've read about great scholars and learned so much from each one. But if I had to name one mathematician who motivated me the most I would have to choose between Archimedes and Euler. Archimedes, one of the greatest thinkers of antiquity, inspired me, awakening my interest in mathematics. He taught me that one must be so consumed by mathematical passion that nothing else matters.

221

Euler, one of the greatest creators and discoverers of mathematics, who kindly, patiently, as he instructed a fairly ignorant princess, taught me from the most fundamental principles of algebra and trigonometry to the newest concepts of infinity and calculus. Euler also taught me that rigorous analysis leads to elegant equations of extraordinary beauty, equations composed of the most simple and superb numbers found in nature. Euler revealed to me the exquisite equation $e^{i\pi} + 1 = 0$ that embodies in its simplicity the most fundamental numbers that rule the mathematical universe.

But my learning is far from complete. I am fully aware that there is much ahead, a myriad of new mathematical truths that I must seek. For I know that I shall use mathematics to draw my vision of the world, the way an artist sketches it.

*Adieu mon ami*—goodbye, my friend. I leave my childhood behind, and now I cross the threshold to embrace my womanhood. This is the beginning of a new chapter in my existence, for I now commence my life as a mathematician. I close the last page of my childhood diary with this theorem that has intrigued me for so long, and that I aspire to prove one day.

Fermat's Last Theorem: A special case says that if $n$ and $2n + 1$ are primes, then $x^n + y^n = z^n$ implies that one of $x, y, z$ is divisible by $n$.

Hence Fermat's Last Theorem splits into two cases:

Case 1: None of $x, y, z$ is divisible by $n$.

Case 2: One and only one of $x, y, z$ is divisible by $n$.

CʒꙄꙄ

# Historical Note

## Marie-Sophie Germain
### (1776–1831)

There was once a mathematical proof so challenging that it stumped scholars for over 300 years. The underlying problem was based on a Diophantine equation found in the classical ancient book *Arithmetica*, written by the father of algebra, Diophantus of Alexandria (c. 200–298 AD). The *Arithmetica* is a collection of problems giving solutions to both determinate and indeterminate algebraic equations, best known on the theory of numbers.

The centuries-old contest started around 1630, when the French mathematician Pierre de Fermat wrote the following as a marginal note in his copy of the *Arithmetica*: "$z^n = x^n + y^n$ has no non-zero integer solutions for $x$, $y$ and $z$ when $n > 2$. *And I have discovered a truly remarkable proof which this margin is too small to contain.*"

Many mathematicians devoted much of their careers to proving what became known as Fermat's Last Theorem (FLT). In 1753, one of the greatest mathematicians of all times, Leonhard Euler, developed a partial solution for $n = 3$ and 4. After Euler, the next major step forward was due to a remarkable woman, the lone French mathematician Marie-Sophie Germain.

Other mathematicians continued expanding the general proof, but it was not until 1994, when British-American mathematician Andrew Wiles of Princeton University developed the complete, elusive proof. Wiles first announced his proof of Fermat's Last Theorem in 1993, but some errors pointed out by Richard Taylor (a former student) showed that proof to be incomplete. A year later, Wiles submitted a complete proof with Taylor's collaboration. Their papers were published together in the 1995 issue of the *Annals of Mathematics*.

Let's go back to the first decades of the 1800s when modern number theory was in its infancy. Carl Friedrich Gauss (1777–1855) was working on analytic functions, called modular forms, which turned out to be important to the new approaches to solving FLT. But Gauss did not produce the sought-after general proof. Sophie Germain did develop a partial proof of Fermat's Last Theorem, competing with other great mathematicians of her time. Her achievement was recorded in 1830. On a footnote in the essay *Théorie des nombres*, published by the French mathematician Adrien-Marie Legendre (1752–1833), he credited Germain with the first general result toward a proof of Fermat's Last Theorem. She used her result to prove Case I of Fermat's Last Theorem for all prime exponents less than 100. This is known today as Germain's Theorem.

Germain, however, never published her theorem, describing it instead in correspondence with Legendre and Johann Carl Friedrich Gauss, as she was not member of a school or group of scholars. Germain's proof was explained by Legendre when he published his own solution for exponent $n = 5$. For over one hundred and seventy years it was assumed that Sophie was a minor player in the collaboration with Legendre. However, a reevaluation of her manuscripts and her correspondence with Legendre and Gauss (*Voice ce que j'ai trouvé*) by contemporary scholars Reinhard Laubenbacher and David Pengelley indicate otherwise.[1]

According to Laubenbacher and Pengelley, "Not only did Germain develop the general version of her theorem independently, but she also deserves credit for substantial additional work

[1] R. Laubenbacher and D. Pengelly, "Mathematical Expeditions: Chronicles by the Explorers." (1998)

previously attributed to Legendre, namely the beginnings of an algorithm for applying her theorem to various exponents."

Aside from Gauss and Dirichlet, Sophie Germain should be considered the principal contributor to number theory in the first quarter of the nineteenth century. And that is only part of Germain's contributions to mathematics.

಩ಌ

*Sophie's Diary* was inspired by Sophie Germain. I wanted to honor Germain and make her known to generations of girls (and others as well), to promote her achievements. Knowing so little about her childhood, I wanted to present a perspective as to how the teenage Sophie must have learned mathematics on her own. Writing *Sophie's Diary* became my way of bringing Sophie to life.

What I wanted most of all was to put into context the environment surrounding the young Sophie Germain who, against all odds, became one of the greatest women mathematicians in history. Long ago, when I fist learned about Germain, I was immensely impressed by a woman, who not only lived during a time of great social turmoil but grew up in an era when women were not permitted in the universities. History can name other great female mathematicians who lived before and accomplished as much, but Sophie Germain did it alone. Hypathia had her father, Theon of Alexandria to teach her; Maria Agnessi had her Rampinelli and other instructors; and Emilie Chatelet had Maupertuis and Clairaut as her tutors of mathematics. Sophie Germain had no teacher.

Equally astonishing it is to learn that Sophie Germain obtained lecture notes from professors at the *École Polytechnique* in 1794. According to her biography, she anonymously submitted her own analysis using a man's name. This intrigued me more, as such daring act would be equivalent to acquiring notes and submitting homework for an advanced university mathematics course, without having attended high school. I wondered how she, as a teenager, could have taught herself mathematics during the siege of the French Revolution. At eighteen, Germain must have known enough mathematics to be so bold and submit her work to not just any professor but one of the greatest mathematicians of the eighteenth century, Joseph Louis Lagrange.

Sophie Germain was born on April 1 of 1776 in Paris, France, during the monarchy of the tragically infamous Louis XVI and Marie Antoinette. Sophie was the daughter of a wealthy merchant. Very little is known about her childhood, but historians define the start of the French Revolution as the time of her mathematical awakening. Germain came of age during a time of social strife in Paris and also one of mathematical revolution. France at that time made an enormous contribution to the fund of knowledge, including the development of modern analysis and mathematical physics.

When she was a teenager, seven of the greatest mathematicians of the eighteenth century worked in the French Academy of Sciences in Paris, not far from her own home: Joseph Louis Lagrange (1736–1813), Gaspard Monge (1746–1818), Pierre Simon Laplace (1749–1827), Adrian Marie Legendre (1752–1833), Johann Friederich Pfaff (1765–1825), Jean Baptiste Joseph Fourier (1768–1830), and Siméon Poisson (1781–1840). These were the great men of science that expanded and further developed the mathematics of Newton, Leibniz, Euler, the Bernoullis, Fermat, and other seventeenth century mathematicians.

Thus, it is tempting to fill in the gaps of Sophie's earlier years with some familiarity of the knowledge available at the time. Also, being born to a fairly wealthy family, one can easily conclude that Germain had access to books to support her study and provide the basis for her own mathematical development. The references to ancient mathematicians and an acquaintance with the works of Fermat, Euler, Newton, Châtelet, and Agnesi in *Sophie's Diary* seem, therefore, justified.

When Germain was twelve, Lagrange published his *Mechanique analytique*, one of the greatest achievements in the analytical method of the calculus and mechanics. Lagrange's work influenced later mathematicians who made the calculus rigorous and used algebra rather than geometry in their arguments. In the years between 1789 and 1793, Lagrange, Legendre, Condorcet and Laplace, among other great scholars, were developing some of their most important work at the French Académie des Sciences in Paris. These were the years I chose to focus on Germain's life and her mathematical development

In 1794, the École Polytechnique opened in Paris, although it was called the École Centrale des Travaux Publics for the first year of its existence. Lagrange was its first professor of analysis, appointed for the school's opening in December 1794. It is not clear how eighteen-year-old Sophie acquired the class notes, but it is suggested it was through a male student enrolled in the school. One can then conclude that Germain knew the student well enough to ask such daring favor. What is recorded in her biography is that Sophie Germain took a man's name, M. Leblanc, and submitted her analysis to Professor Lagrange. In *Sophie's Diary*, it became imperative to introduce a character named Auguste LeBlanc with whom Sophie has some interaction, mainly to satisfy her eagerness to learn more mathematics.

Lagrange's mathematical analysis was rigorous and difficult, even for the students attending his lectures at the École. Thus, the professor must have been curious to know who this Monsieur Leblanc was. Eventually, Lagrange discovered that M. LeBlanc was a young woman. Impressed by her brilliance and resourcefulness, Lagrange became her mathematical counselor.

Sophie also had a professional relationship with other professors at the *École* and, without attending lectures, she expanded her initial analysis. Her education was, however, disorganized and haphazard, since she never received the rigorous education that she eagerly desired. Knowing this, perhaps, Sophie sought out advice from the greatest mathematicians of her time and was courageous enough to submit her own ideas and solutions to very difficult mathematical problems.

Sophie wrote to Adrien-Marie Legendre, another of the great French mathematicians, professor at the École and colleague of Lagrange. In her correspondence, Sophie addressed some problems from his 1798 book titled *Essai sur le théorie des nombres* (*Essay on the Theory of Numbers*). By then, Legendre must have known that she was a woman and must also have seen the genius in her work. He corresponded and collaborated with Sophie for a number of years and included some of her mathematical discoveries in a supplement to the second edition of his *Théorie*.

Sophie was undoubtedly impressed by the work of German mathematician Carl Frierich Gauss (1777–1855). In 1801, Gauss published his masterpiece on the theory of numbers, *Disquisitiones arithmeticae* (*Arithmetical Inquisitions*). Three years later, a twenty-five-year-old Sophie began to correspond with Gauss, sending him some of her own mathematical analysis. How did she develop the audacity to write to him? The only plausible answer is that Sophie had developed a thorough understanding of the methods presented in Gauss' dissertation. In addition, seeking acceptance as a genuine mathematician, what better way than go to the prince of mathematics himself.

Between 1804 and 1809, Sophie wrote many letters to Gauss, initially adopting again the pseudonym "M. LeBlanc" because she feared Gauss would ignore her letters if he knew she was a woman. In these letters she sent proofs of number theory, and Gauss praised her ingenuity and mathematical ability.

In fact, Gauss did not know she was a woman almost his own age until after the French occupation of his hometown in Germany in 1806. Sophie contacted a French commander who was a friend of her family and asked him to inquire about the well being of Monsieur Gauss. Eventually Gauss discovered that "M. LeBlanc" was in reality Sophie Germain. He was truly impressed and even mentioned her mathematical proofs and insight in a letter Gauss wrote to his colleague Heinrich Wilhelm Matthäus Olbers (1758–1840). The correspondence with Germain ended in 1809 when Gauss stopped replying to her letters.

Sophie continued working alone, most notably in number theory, resulting in her partial solution to Fermat's Last Theorem, which became known as "Germain's Theorem." Sophie demonstrated the impossibility of solving the equation $x^n + y^n = z^n$, if $x, y, z$ are not divisible by an odd prime $n$. This remained the most important contribution related to Fermat's Last Theorem (1738) until the next result contributed by mathematician Kummer in 1840.

In addition to number theory, Sophie was attracted to mathematical physics. In 1808, the German physicist Ernst F. F. Chladni came to Paris and drew the attention of the scientific community with his experiments on vibrating plates. He sprinkled sand on elastic surfaces, strummed the edges with a bow, and noted the resulting patterns, exhibiting the so-called Chladni figures. The Institut de France set a prize competition with the following challenge: *formulate a mathematical theory of elastic surfaces and indicate just how it agrees with empirical evidence.* A deadline of two years was given for mathematicians to submit such mathematical theory.

The two-dimensional elasticity theory was considered too formidable for most mathematicians and, since Lagrange thought that the mathematical methods available were inadequate, most scientists did not attempt to formulate it. Sophie, however, accepted the challenge and spent the next two years trying to derive a theory of elasticity. In 1811, she was the only entrant in the contest, but her work did not win the award.

Her historians conclude that Germain had not derived her hypothesis from physical principles, and her analysis lacked the necessary rigour owing to her deficiency in analysis and the calculus of variations. Sophie's essay did generate more interest in the topic and provided needed insight to pursue the theory. Lagrange, who was one of the judges in the contest, amended Sophie's calculations and developed an equation that could better describe Chladni's experiments.

Without a winner, the contest deadline was extended by two years. Once more Sophie submitted the only entry in which she demonstrated that Lagrange's equation yielded Chladni's patterns in several cases, but she could not give a satisfactory derivation of the equation. Nevertheless, the panel of judges, which included the best mathematicians of her time, deemed Sophie's second mathematical memoir worthy of an honourable mention.

Once again the contest was re-opened and, in 1815, Sophie won the *grand prix*, a medal of one kilogram of gold grand prize. What should have been Sophie's greatest achievement and source of pride, however, turned into a bittersweet victory. She received a laconic response from Siméon Denis Poisson, one of the judges and chief rival on the theory of elasticity, who wrote that her analysis still contained deficiencies and lacked mathematical rigour. Sophie did not appear at the award ceremony, to the disappointment of many who wished to meet her. It was suggested[2] that she thought the judges did not fully appreciate her work, and that the scientific community did not show the respect that seemed due to her. In fact,

---

[2] L. L. Bucciarelli and N. Dworsky, *Sophie Germain: An Essay in the History of the Theory of Elasticity* (Dordrecht - Boston, Mass., 1980).

strangely enough, it was reported that Poisson avoided any serious discussion with Sophie and ignored her in public.

It is easy to imagine Sophie's disappointment at the unwelcoming response from the scholars she so desperately sought out. Although she was the first to attempt solving such a challenging problem, and others used her analysis in elasticity to derive their own results, Sophie was not taken as seriously as she deserved to be.

Sophie extended her research and, in 1825, she submitted a paper to a commission of the Institut de France, whose members included Poisson, Gaspard de Prony and Laplace. Sadly, the commission ignored Sophie's essay and her work went unrecognized until 1880, when it was found among de Prony's papers.

Joseph Fourier, another distinguished French mathematician who clearly admired Sophie, befriended her and tried to help. In 1823, he wrote to invite her to the meetings at the Academie des Sciences. She also earned the respect of many others. French physicist Jean-Baptiste Biot wrote that "Mlle Germain is probably the one of her sex who has most deeply penetrated the science of mathematics, not excepting Mme du Châtelet, *for there was no Clairaut*" (Emilie Châtelet's mathematical tutor)[3].

Nevertheless, Sophie Germain persevered, working alone until her death. Sophie was stricken with breast cancer in 1829 but, undeterred by her illness and the revolution that shook Paris again in 1830, she wrote papers on number theory and on the curvature of surfaces. She outlined a philosophical essay, which her nephew published posthumously as *Considérations générale sur l'état des sciences et des lettres* in the *Oeuvres philosophiques*.

Marie-Sophie Germain died in Paris on June 27, 1831. Years later, a plaque was erected on the house she died, naming Sophie Germain *philosophe et mathématicienne*. The rather unpretentious building where Sophie lived throughout her adulthood is still standing on 13 Rue de Savoi, across from Seine River, and a few blocks from Rue St Denis where she grew up. The plaque on the upper right side of the arched entrance to the apartment building states laconically: SOPHIE GERMAIN, PHILOSOPHE ET MATHEMATICIENNE, NEE A PARIS EN 1776 EST MORTE DANS CETTE MAISON LE 27 JUIN 1831.

---

[3] *Journal de Savants*, March 1817.

I wrote this book to honor the memory of Sophie Germain, outstanding woman, philosopher, and brilliant mathematician who became immortal thanks to her work with prime numbers, her partial solution to Fermat's Last Theorem, to her work in mathematical physics, and her contribution to the study of elasticity.

Sophie Germain's Theorem:

Let $n$ be an odd prime. If there is an auxiliary prime $p$ with the properties that
Case 1: $x^n + y^n + z^n = 0$ mod p implies that $x = 0$ mod $p$, or $y = 0$ mod $p$, or $z = 0$ mod $p$, and

Case 2: $x^n = n$ mod $p$ is impossible for any value of $x$, then Case I of Fermat's Last Theorem is true for $n$.

To see why Case I holds for a Germain prime, suppose $p = 2n + 1$ is a prime, where $n$ is an odd prime. Then for any number $0 < a < p$, Fermat's Theorem implies that $(a^n)^2 = a^{p-1} = 1$ mod $p$. Therefore $(a^n - 1)(a^n + 1) = 0$ mod $p$, and since $p$ is a prime, we must have either $a^n = 1$ mod $p$ or $a^n = -1$ mod $p$. This means that if $x$, $y$, and $z$ are not congruent to 0 mod $p$, then $x^n + y^n + z^n = \pm 1 \pm 1 \pm 1$, which can never equal 0 mod $p$. Hence property (1) in Sophie Germain's theorem is true. Moreover, it is impossible for $x^n = n$ mod $p$ to have a solution, establishing property (2) in Sophie Germain's theorem.

For each odd prime $n < 100$, Germain gave a prime $p$ for which her theorem applies, thereby showing that case I of Fermat's last theorem holds for all prime exponents less than 100.

Legendre generalized Germain's argument to show that properties (1) and (2) hold for the odd prime exponent $n$ provided that one of the numbers $4n + 1$, $8n + 1$, $10n + 1$, $14n + 1$, or $16n + 1$ is a prime. When $n = 3$, the numbers $4 \cdot 3 + 1 = 13$, $10 \cdot 3 + 1 = 31$, and $14 \cdot 3 + 1 = 43$ are all prime, but only $p = 13$ satisfies properties (1) and (2) for $n = 3$ ($p = 2 \cdot 3 + 1 = 7$ also satisfies the two properties.) With this result Legendre could show that all prime exponents less than 197 satisfy Case I of Fermat's Last Theorem. Germain and Legendre stopped at 197. The first prime that works for $n = 197$ is $p = 7487 = 38 \cdot 197 + 1$.

# Marie-Sophie Germain

**Achievement Areas:** Number theory and mathematical physics

**Major Contribution:** Germain prime numbers. She is also one of the founders of mathematical physics.

**Timeline:**

| | |
|---|---|
| 1776 | Marie-Sophie Germain is born April 1 in Paris, France. |
| 1789 | Discovers passion for mathematics after reading Archimedes story. French Revolution begins with the fall of the Bastille. |
| 1794 | Collects lecture notes from a student at the École Polytechnique and submits a paper to Professor Joseph-Louis Lagrange using the pseudonym "M. LeBlanc." |
| 1804 | Writes to German mathematician Carl Friedrich Gauss concerning his essay *Disquisitiones Arithmeticae* (1801) and submits her own analysis signing the letter as "M. LeBlanc." |
| 1811 | Submits a theoretical paper to the French Académie des Sciences competition, explaining the vibration patterns demonstrated by Ernest Chladni. |
| 1813 | Académies of Sciences awards her honorable mention for her second entry on Chladni's vibration plates. |
| 1816 | Wins the grand prize of the Académie des Sciences for the best paper on her mathematical theory of vibrations of general curved and plane elastic surfaces. |
| 1831 | Gauss nominates Sophie for an honorable doctorate at the University of Göttingen in Germany. She did not live to receive it. |
| 1831 | Sophie Germain dies in her Paris home on June 27. |

## Suggested Reading

1. L. L. Bucciarelli and N. Dworsky, *Sophie Germain: An Essay in the History of the Theory of Elasticity* (Dordrecht - Boston, Mass., 1980).
2. Biography in *Biographical Encyclopedia of Mathematicians*, Editor: Donald R. Franceschetti, Marshall Cavendish (New York, 1999).
3. Biography in *Dictionary of Scientific Biography* (New York 1970-1990).
4. Biography in *Encyclopedia Britannica*.
5. Margaret Alic, *Hypatia's Heritage*
6. A. D. Dalmedico, "Sophie Germain," *Scientific American* 265 (1991), 117-122.
7. L. S. Grinstein and P. J. Campbell (eds.), "Women of Mathematics," (Westport, Conn., 1987), 47-56.
8. Lynn M. Osen, *Women in Mathematics* (The MIT Press, Cambridge, MA 1995).

## Acknowledgements

*I am indebted to the following people and organizations for allowing me to use their wonderful material to illustrate this book: Professor L.L. Bucciareli, The Pierpont Morgan Library in New York, the Perry Castañeda Library Map Collection of the University of Texas at Austin, and to the Foundation E.G. Bührle Collection in Zurich (Switzerland).*

## List of Illustrations

Front Cover: Camille Corot painting, *A Girl Reading*, c.1845/1850 – Courtesy of the Foundation E.G. Bührle Collection, Zurich (Switzerland).

Map of Paris in 1789 – Courtesy of Perry Castañeda Library Map Collection, The University of Texas at Austin.

Hôtel de Ville in Paris – Engraving from *A Pictorial History of the World's Great Nations from the Earliest Dates to the Present*

236

*Time*, by Charlotte M. Yonge, published by Selmar Hesse, NY, 1882.

The siege of the Bastille, July 14, 1789 – The Pierpont Morgan Library, New York. PML 140205 #16.

The arrival of the royal family in Paris after the assault in Versailles, October 6, 1789 – The Pierpont Morgan Library, New York. PML 140205 #31.

The attempted demolition of the Vincennes castle by the angry mob, February 28, 1791 – The Pierpont Morgan Library, New York. PML 140205 #48.

The arrival of the royal family in Paris after their failed attempt to flee the country, June 25, 1791 – The Pierpont Morgan Library, New York. PML 140205 #54.

The storming of the Tuileries Palace, August 10, 1792 – The Pierpont Morgan Library, New York. PML 140205 #67.

Attack on the Palais-Royal – Engraving from *A Pictorial History of the World's Great Nations from the Earliest Dates to the Present Time*, by Charlotte M. Yonge, published by Selmar Hesse, NY, 1882.

The royal family is taken to be imprisoned in the Temple, August 13, 1792 – The Pierpont Morgan Library, New York. PML 140205 #69.

The massacres of the prisoners, September 2-5, 1792 – The Pierpont Morgan Library, New York. PML 140205 #72.

Sophie Germain, a sketch from Stupuy, reproduced from L.L. Bucciarelli's book, Reference 1.

**About the Author**

Dora Elia González y Musielak pursued advanced degrees in aerospace engineering stimulated by her love of mathematics. In addition to her research in space rocket propulsion, Dr. Musielak teaches science and mathematics courses at the University of Texas in Arlington.

Made in the USA
Lexington, KY
26 March 2016